The Basics of Print Production

2nd Edition

D1466031

The Basics of Print Production

2nd Edition

by
Mary Hardesty Kuhn

Printing Industries Press
PITTSBURGH

Printing Industries Press
Printing Industries of America
301 Brush Creek Road
Warrendale, PA 15086
Phone: 412-259-1770
Toll Free: 1-800-910-4283
Fax: 412-741-2311
Email: membercentral@printing.org
Internet: www.printing.org/store

Table of Contents

This book is dedicated to my children.

Special Thanks
There are far too many people to whom I owe my thanks to list them all, but I want to extend a special thanks my vendors and suppliers who have helped to stay abreast in this ever changing industry.

Preface

As the title indicates, the purpose of this book is to provide a basic overview of the production process—a broad general view of the steps required to take a creative concept to a printed piece.

The information provided in this book is not an in-depth study. I have only touched on the high spots in each section and then in only very broad terms. The goal is to enlighten and inform. This text is not to be used as a comprehensive study of the production process. That would require a series of books on each of the sections highlighted in this text.

I hope that this text will provide the reader with a general awareness of the intricacies of the printing process and an appreciation for the number of steps and amount of planning required to get to the final product.

As with anything, having a basic understanding of the production process promotes better communication. Being aware of the basics of the process helps anyone who is involved even peripherally—whether client or vendor, from account service to billing clerks, from creative to management—recognize and appreciate the time and cost factors as well as the influence of each decision or step on the overall process.

Acknowledgments

Years ago, I was told, "The debt owed for knowledge shared is to pass it on." I offer this text with my sincere thanks to all of the men and women in the Graphic Arts Industry who have shared and continue to share their knowledge and expertise.

Introduction

The illustration on the next page depicts the major steps in the production process, from approval of the creative concept and copy, through the final printed pieces packaged and ready to ship. Within each step are a variety of processes. Therefore, determining the best process for a project requires specific information about the job. Obtaining the information needed to plan, estimate, and produce a print job is an integral part of the print production process. We refer to this information as *job specifications.* The specific information and how it is communicated can be as important to understanding print production as learning how a printing press works.

Learning the basics of the process includes becoming familiar with how each piece of information in a specification can impact the choice of processes used to complete a printed product.

Print production is all about communication, and everyone in production acknowledges that there is no such thing as too much information.

In addition to the set of specifications about the job, it is always desirable to have a printout or hard copy that shows at least the direction of the job, if not the approved design. Failing that, line drawings or sketches can be used to communicate the shape and structure of the piece.

Information Required for Planning a Printed Piece

In addition to the budget allowed for the printed piece, the information required for planning includes the following:

1. **Quantity.** The total number of finished pieces that you want to produce. The quantity often influences the choice of printing method, paper stock, size, number of colors, and the finishing processes used.

   If no specific quantity has been determined, then a range of quantities can be used. If there is a wide range of quantities, it may be necessary to plan the job in different ways for the high- and low-end quantities.

Steps in the Process
from Approved Layouts to the Delivered Printed Piece

Submit specifications for quote and timeline

Review specifications with layouts

Submit quote with specifications and timeline to client
for budget approval

Perform scanning, color correction, retouching, cropping,
system work, loose color

Review layout files and keylines with quote

Finalize quote and timeline

Receive signed estimate and/or P.O. from client

Build final print-ready files with high-resolution images

Preflight file

Trap and impose files for press

Produce composite digital proofs or hard copies

RIP files to imagesetter for film output*

Create final composite proofs for client approval*

Make printing plates

Perform press makeready

Print job on press

Perform binding and/or finishing operations

Package and fulfill job

Ship and deliver job

*Workflows using computer-to-plate systems, direct-imaging presses, and digital presses have variations on these steps (see sections on proofing and printing).

2. **Print area and finished size of the piece.** In order to determine the materials that will be required, as well as the size of the equipment on which it will be printed, you must know the flat (unfolded) size of the piece, whether the printing extends to its very edge, and the size that it will be when finished. This information helps to determine the size of the film, paper, press, and in some cases the finishing equipment that will be required to produce the piece. The size and folding/binding requirements may dictate that the grain of the paper run in a particular direction, e.g., for ease of folding.

3. **Number of colors.** Another planning consideration is the number of colors to be used. Is it a process-color job, which requires four inks (cyan, magenta, yellow, and black)? Are PMS and match colors required? Will the job be varnished or coated? All of these considerations affect the number of plates and printing units required to produce the job.

   In addition, this information can influence the proofing method selected, especially if there are match colors or critical color involved, as opposed to straight four-color process. The number of colors and the way they interact can also help to determine if multiple press passes will be required to achieve the desired result. In some cases the printing method is determined by the number and types of colors required.

4. **Types of images and art.** The types and quality of these elements need to be considered in selecting the paper. Both the quality and finish of the paper affect the way the art and images will reproduce. The type of images and art may dictate the printing and proofing methods used, the screen ruling of halftone images, and perhaps even the estimated printing speed. The way that the creative elements are arranged can impact the bindery and finishing methods selected; for instance, some layouts may require expensive hand bindery operations instead of mechanical operations.

   The estimated prepress costs for a project will be based, in part, on the number of downloads and/or scans, the amount of system work that will be needed. as well as the number and types of loose color proofs you expect to need to get final approval. How the files are furnished and the number of rounds of changes required to get final approval will also affect the costs.

5. **Paper or other printing substrate.** The substrate is an integral element of the job and contributes greatly to the cost of producing a printed piece. It is important to get as much information about the substrate to be used as possible. Because it is such a

critical factor in planning a print job, the stock to be used starts to impact the process in prepress and continues through shipping.

In prepress, for example, the material on which an image will print impacts the color separations, balance, and density, and also a consideration in determining the dots per inch (dpi). Not only are the finish and quality of the substrate important, but the weight or thickness can also be a concern because some equipment has limits with respect to the weights of stocks on which they can print. Paper thickness, along with the design, will often be a factor in determining the finishing and/or binding methods used. In addition, paper weight is a concern for any material that will be mailed. For example, using a slightly lighter paper could result in savings in postage or shipping costs, but if it is too light, it may not meet the postal requirements to mail.

6. **Bindery and finishing.** This is a collective term referring to the postpress processes required to convert the flat printed press sheet to the final size and shape of the printed piece. Some finishing operations (e.g., folding and even simple diecutting) can be done in line on specially equipped presses, but most postpress operations are performed offline in the bindery. These processes include trimming, diecutting, embossing, gluing, and folding. The layout of the printed piece on the press sheet must be matched to the capabilities of the binding and finishing equipment.

7. **Fulfillment and packaging.** Fulfillment and the packaging of the materials in a specific manner also must be considered during the planning process. The need to protect the materials, the size and shape of the boxes and/or shipping cartons, and the amount of hand work impact both the cost and time requirements. Fulfillment can sometimes be such a large part of the project that it will even be a factor in choosing a vendor. For instance, if the preference were for a single vendor to be responsible for the entire project through shipping, then only vendors with fulfillment capabilities would be considered.

8. **Freight.** Freight, the actual costs of shipping the materials, can be a critical part of an estimate. Freight charges should be included in the client's project budget to avoid being over budget on the final costs. If there is not enough information to estimate freight, it should be noted prominently on the estimate that the estimate or quote is without freight. The estimate will usually state that the prices are FOB (free on board) the vendor's dock.

As soon as possible, the freight costs should be determined and submitted in writing. Airfreight to make a delivery date can sometimes be more expensive than the overtime and rush charges needed to make a ship date, and it should not be an element to be ignored.

As you can already see, this information is interactive. Each of these factors can impact, or in some cases even dictate, others which is why it is necessary to understand the basics of each process and the effect of each choice on the whole.

The goal of this book is to explain the process of producing a printed piece, starting with the design decisions and proceeding through shipping, examining the choices and their impact at each step in the process. Communications and understanding are the only way to ensure that the selections and decisions made on the individual elements (e.g., paper, image treatment, screen ruling, printing method, finishing, and so on) will result in the desired look and feel of the finished piece while staying within the budget and timeline of a project.

Tracking Costs and Timelines

All of the elements and processes required to produce a project impact one another. Any change—whether it is a quantity or an image element—not only affects the estimate but also alters the timeline.

Tracking costs by doing an adjusted estimate or additional estimate for the costs, along with a revised timeline as change requests are made, is the only way to track and document the discrepancies between the original quote and the final invoice. Using revised estimates and timelines to confirm change requests will help to clarify the impact of the changes and keep everyone in the loop on costs and time.

Getting sign-offs as changes occur not only presents an opportunity to evaluate decisions and change requests before additional costs have been incurred but also eliminates the need to reconstruct the final costs for billing. Tracking costs throughout a project eliminates surprises when the job is invoiced by building an ongoing set of documentation on the job costs, making the billing process more timely and accurate. In any project, it can be difficult if not impossible to reconstruct all the changes or additions to justify additional costs after a project has been completed.

Tracking the impact of changes and alterations to the schedule can be as important as tracking the impact to costs. Staying on top of the timeline with periodic status checks can mean the difference between making a delivery date, staying in budget, or both. Knowing immediately that a job is off schedule gives us the opportunity to investigate the options available to get it back on track. Failing that, it would at least be possible to alert the concerned parties to the probability of a missed delivery in advance of the date, as it is always preferable to know that material may be delayed before the day it is due to deliver.

1
Specifications

Specifications, or "specs," are a description of the piece to be produced. They include information about the materials and processes, the overall size of the final product, the type of material on which it is to be produced, the number and type of images or illustrations, the number of colors (four-color process, PMS, and/or match colors), bleed, finishing information, and the quantity to be produced. Specifications are used to calculate the cost of a project. Depending on whether the specifications are preliminary or final, the costs would be presented as an estimate or a quote.

Specifications are also used to create a *timeline,* a schedule or projection in days of the amount of time that will be required to produce the materials. Every set of specifications should include the date the information was released for estimate or quote, the date the estimate is due to be considered, the projected release date of the materials to the vendor, and the date it must arrive or be ready to ship.

There are two types of specifications: preliminary and final. Both types contain the same basic information. The difference is that there is less detail in *preliminary specifications* because the final design has not been done. Therefore, it is necessary to make assumptions or educated guesses about some of the elements. Generally, preliminary specifications are used to generate estimates, or a range of costs for a project, and are subject to review with the final art and specifications.

Final specifications should be written from the final approved layouts. The information should be detailed and specific. Marked-up copies of the approved layouts should be attached to the specification sheets. Final specifications are used to generate quotes. As long as the project specs remain as stated in the preliminary specifications, the costs should not change.

Final specifications need to be as detailed as possible and often are not done until the piece is released to a vendor. Using preliminary estimates to select a vendor is not uncommon, but the selected vendor will need to generate a final quote once the final specs are written.

Information for Specification Sheets

- Date issued
- Date required
- Projected date that materials will be released
- Shipping/delivery date
- Job reference/number
- Element (e.g., brochure, business reply card, envelope, header, shelf talker, necker)
- Quantities or range of quantities for each element
- Prepress and color information
- Number of images required for each piece: illustrations, photography, etc.
- Number of crops and/or re-sizes for a single image—sometimes one scan will be used in various sizes and may require handling more than once
- System work—retouching, merging, color corrections, etc. (this cost will be based on history with the client as a rule or left out of the estimate and specifically noted as an indeterminable cost)
- Printing/finishing information
- Maximum overall size of piece flat and finished
- Number of colors, e.g., four-color process and/or flat, custom, or match; coating; etc.
- Printed one side or two
- Page count or panels where applicable
- Art/image information—type, number, size, etc.
- Format in which files will be submitted—layout or mechanical, Quark, InDesign ...
- Type of paper or substrate—weight and quality
- Print area, amount of coverage, bleeds
- Finishing requirements—e.g., diecutting, foil stamping
- Bindery requirements—e.g., saddle stitching, perfect binding, GBC, Wire-o
- Special instructions—information or instructions about the piece that are not covered in the other sections
- Packaging requirements (e.g., shrink wrap, inserted in envelope, kits, bulk packed)
- Freight

The purchase order (PO) should be written to the quote, not the estimate. Very often that is not the case, but in a perfect world that is how it would work.

Contents of a Set of Specifications

Once there is a creative layout, it is possible to begin to set the parameters of the specifications. A set of specifications has several sections. Those sections and general information are the same for preliminary or final specifications, although the final specifications require more detailed information within each section. In most cases, the final specifications are built from the preliminary specs.

Since there is less emphasis on the specifics and more of a need for worst case and general information in preliminary specs, the information is generally determined by an initial layout. This is acceptable in most cases because the information is being used to generate a range of costs as opposed to a hard quote. The layout can be used to determine the approximate flat and finished size of the piece, as well as how the creative person visualizes the concept, the number of images, the number of colors, the coverage, the complexity of the design, and the finishing and diecutting information.

Prepress Section

Information in the prepress section normally includes the following:
- Number of images/elements
- Final size of each image
- Source of the images
- Raw or converted
- Amount of retouching
- Line screen (halftone screen ruling)
- Number and type of loose proofs (color-corrected visuals of the individual images) required to get the color and retouching approved
- Specific instructions/directions regarding system work or assembly
- Instructions on delivery of the final approved files

In the preliminary specifications, the number of images should reflect the maximum number likely to be used and the most likely source. Since there are no specific parameters regarding the amount of retouching required (e.g., 30 min./image, the number of PDF and/or analog proofs per image, and so forth), it is necessary to set parameters that, at least, will provide an estimate of the time and materials required. If competitive estimates are being solicited from a number of vendors, it is impossible to compare the estimates equally without specific parameters.

Doing a set of final prepress specifications usually requires the selection of a print vendor because the choice of printer can influence or determine a number of the elements in the set of final prepress specifications: film and proofs, the press gain, maximum printing densities, and line screen.

Printing/Finishing Specifications Section

The printing/finishing specifications section includes the composite proofs, plates, printing, and finishing requirements for the project. Working from the layout, the preliminary or general guidelines for the final production can be determined. While not final, these directions are generally accurate unless the design direction changes radically. This section includes the following specific information:

- Size information
- Standards of weight and quality of the material on which the piece will be printed
- Whether there will be any bleeds
- Number of colors
- Image content if the printing vendor will not be doing the prepress estimate
- Configurations of the final piece, like the page count or number of folds
- Finishing and bindery processes required to produce the finished piece

The last piece of information required for this section is the quantity. When building the preliminary specifications, the final quantity may not have been determined yet. In that case, the vendor will be asked to estimate the costs for a number of quantities.

Special Instructions Section

Some projects require more than just the usual printing and finishing operations. The instructions regarding those requirements should be listed under the special instructions section. For example, a brochure layout indicates that a piece of ribbon will need to be tied around each finished book. The special instructions would include the width, color, and fabric type of the ribbon, as well as instructions on how the ends should be cut and the type of bow or knot. Any time a project includes an element that is not part of the printing specifications, it should be outlined as a separate set of instructions with samples requested.

Packaging Section

The packaging section contains the information about how the material is to be packed. There are a variety of ways in which materials can be packaged. For some projects, the finished product is shrink-wrapped. Shrink-wrap is clear plastic that is loosely wrapped around print material and then placed in a heated tunnel that "shrinks" the plastic to conform to the printed material. Other packaging options can be as simple as using paper bands or rubber bands to keep the product from shifting during transit. Another option would be to bulk-pack the finished product in cartons stacked on skids.

Some projects include a number of elements that need to be packed into kits. **Kits** are boxes that contain a predetermined number of each of several elements, or a variety of kits that have different quantities and components depending on the style of the kit. These require custom boxes and a great deal of handwork. They can also take up a large amount of room. Specialty vendors called fulfillment centers can be hired to do the kit building, but if the information is not included in the specifications, the costs associated with kit building will not be included in the estimate.

Some projects require that elements be packed more than once. For instance, the material needs to be shrink-wrapped into packages and then packed in cartons, or first packed into cartons and then stacked on skids.

It is important to include all of the information about how the material is to be packed in the specifications:
- Type of material
- Number per unit
- Whether the units are to be repacked in specific quantities, such as twelve units to a carton
- Whether there are limits of weight, height, or width that the packaging must conform to
- Any other specific packing and labeling information that the project requires

If the information is not available when the preliminary specifications are written, include a note or disclaimer that indicates what the estimate includes with regard to the handling of the finished materials, and for how long they will be stored before additional costs would be incurred for the storage.

Shipping and Handling

Each box or skid must be prepared for shipment. The method of shipping—as well as the number of drop shipments and the final destinations—are all factors that impact the costs of the freight and thus the project budget.

Freight and shipping can even have an impact on production. Limitations on sizes and weight for some types of freight can dictate the maximum size of a package. For example, a large piece might have to be printed to fold three times instead of twice. It is important to include any shipping information that you have in the specifications from the start.

Freight is the method by which the material is actually moved from the vendor to its destination. Freight charges vary widely. Along with the regulations and services that each company offers, the costs are directly related to two factors: the weight of the shipment and time allowed for the material to reach its final destination. If freight cannot be estimated in the preliminary estimate, it is imperative that all parties be aware that the estimate is FOB and that the freight estimate will be added to the budget as soon as it is possible. If the material can not be produced in time to use normal freight and must be air-freighted to make the delivery, the freight costs can be triple or more.

Why Do Preliminary Specifications?

Since we have stated that final specifications need to be done when all the specific information about the project has been determined to get an accurate quote, the logical next question is "Why do preliminary specifications at all?"

Preliminary specifications and timelines are used for a number of purposes. If the preliminary specifications are not likely to change a great deal, it is possible to use them to prequalify vendors so that when the final specifications are done, the vendor selection process will have already been completed and only the selected vendors need do final quotes.

1. If no budget or set delivery time has been established for a project, both a budget and time requirement can be determined and submitted for approval prior to final creative based on the preliminary specifications.

2. If a concept requires a lot of hand work or is an involved display, preliminary specifications can be used to generate a constructed comp, a hand-made blank working "model" of the piece. The comp is built using materials of the weight and quality required for the piece. A comp should be made in the size and finished configuration indicated by the layout. Having blank comps, budgetary estimates, and timelines help to determine the feasibility of a design or concept. If the concept cannot be constructed or the cost is prohibitive, it would be best to determine that early in the process— not when the piece is on deadline.

3. Preliminary specifications can be used when a project has a fixed budget and/or delivery date to determine if the concept can be produced within the budget and timeline. Using a set of preliminary specifications, a vendor can help to determine the viability of a concept within a client's stated constraints before the project goes

into final creative. If the concept is too expensive or requires more time than has been allowed, the decision can be made to change the concept, expand the budget, move the delivery date, or scrape the project altogether before an excessive amount of time and money have been committed to it in the creative process.

4. Short-run or small-quantity projects that have time and budget constraints sometimes make it necessary to request estimates from vendors representing different production options; e.g., a screen printer and a digital output vendor. The preliminary specifications can be used to solicit estimates and printed samples of pieces with similar style and ink coverage on the same stock, making an advised comparison of cost, time, and quality possible. Since the choice of processes might have an impact on the design, it would be best to make the process determination as early in the process as possible. It is also necessary to know which process will be used to do the final files since the requirements (e.g., dot gain and screen rulings) are different.

5. If a project has a critical delivery date that cannot be moved, doing preliminary specifications will help to generate a project timeline by working backward from the shipping date determined by the delivery date. By including the delivery information in the preliminary specifications, the vendors can build a timeline that takes the availability of equipment, the time needed to order materials, and so forth into consideration and determine whether it is possible to produce it in the time frame. The preliminary specifications also identify when the final approved materials need to be in their hands.

A request for estimate on a piece with a fixed delivery date should always include a statement stressing that if a vendor can not commit to the delivery schedule as stated, the vendor should either decline to bid or furnish a timeline that indicates the latest date by which the vendor could receive the materials and, as a separate line item, the amount of overtime and/or rush charges that would be required. Any time that overtime or rush charges are required, it is best to make them a separate line item so that the involved parties can see the impact that making the delivery will have on the costs of the project and have the opportunity to weigh all their options.

Timelines

A timeline is the estimated amount of time it will take to move a project from one specified point in the process to another. Just as the specifications create the general parameters to estimate the cost of a project, they can also be used as a guide to determine the approximate amount of time the production processes will take. This type of time-line is called a ***production timeline,*** since it is used to map out the time requirements for each step in the production process. When a project is being handled by only one vendor, or when a single vendor is responsible for the entire project and will coordinate all production processes, that vendor can build the timeline. If multiple vendors will be involved, all vendors must submit time requirements for their processes, and an overall timeline will need to be developed from the input of the individual vendors. A timeline can be formatted to indi-cate the number of days allowed for each step in the process or be built in a calendar format with specific dates. The format is usually dictated by the status of the project when the timeline is built.

A timeline can be built to generate different information. In some instances, it is necessary to determine the amount of time that will be required to get a project completed from the initial concept through final delivery. These are ***project timelines.***

Building a full project timeline is a team effort. There must be input from the account team and/or client, the creative group, the mechanical department, and all involved vendors. These timelines are generally used by the agency or client to make sure that the project stays on track and to raise red flags when things begin to fall behind.

Samples of Specifications, Budgets, and Timelines

The remainder of this chapter contains a number of samples showing two requests for estimates using preliminary specifications (**Figure 1-1** and **Figure 1-2**), a preliminary budget and timeline (**Figure 1-3**), a request for preliminary budget and timeline for POP and POS mate-rials (**Figure 1-4**), and a request for quote for loose color (**Figure 1-5**).

REQUEST FOR ESTIMATE
Preliminary Specifications

Date: June 10, 2013

Estimated date due: June 12, 2013

Date to be released to vendor: Not determined

Delivery/ship date: Not determined

From: Production Manager
Phone: 999-555-0000, ext. 1222
Email address: pmanager@agency.com

Alternate contact: Production Assistant
Phone: 999-555-0000, ext. 1111

Job reference: Brand X New Brochures (2)

PREPRESS
6 images: Raw format files, posted to FTP
       3 furnished 4×5-in. transparencies
       3 furnished 35-mm slides

Images
Download image files, crop and scale to finished size of 4×2 in. at 350 dpi

Retouching
Estimate 30 min./image for retouching and system work.

Color correction and proofs
2 rounds of digital loose color proofs. Agency will use 72-dpi images with the same names For Position Only (FPO) in mechanical files, to be updated and replaced with final approved high-res images at final.

Scheduling information
Please indicate number of days needed from receipt of images to final approval. For purposes of the timeline, please allow one full working day from receipt of proofs for approvals.

Figure 1-1. A sample request for estimate to have six images scanned, retouched, and proofed based on preliminary specifications.

REQUEST FOR ESTIMATE
Preliminary Specifications

Date: June 10, 2013

Estimated date due: June 12, 2013

Date to be released to vendor: Not determined

Delivery/ship date: Not determined

From: Production Manager
Phone: 999-555-0000, ext. 1222
Email address: pmanager@agency.com

Alternate contact: Production Assistant
Phone: 999-555-0000, ext. 1111

Job reference: Brand X New Brochures (2)

FURNISHED FILES

Mechanicals for each of two elements will be furnished on disk/FTP with approved high-res final files and approved color proofs.

Vendor to adjust files to press as needed, build press forms, and trap. Vendor to furnish full-form imposition proofs, constructed one-color proofs, and ink drawdown of PMS on production stock for final approval prior to plating.

Approved furnished loose proofs to be used at press for color; digital proofs for content and position only. Marked-up color lasers of all elements will be furnished with files. If furnished on disks, they will have print window and list of software and the version numbers in the files.

ELEMENT #1—6-PANEL BROCHURE

Quantities: 1,000 2,500 5,000 10,000

Flat size: 12¾ in. + bleed × 6¼ in. + bleed   **Finished size:** 4¼×6¼ in.

Prints as 4-color process + 1 PMS + overall coating 2 sides

Coverage: Full bleed, heavy coverage, product color match required

Paper: Lustro Gloss Enamel Cover 100#

Print 2 sides, **score** in thirds, **Z-fold,** and **trim** to final size

Press check: Customer press check for color approval, both sides

Figure 1-2. A sample request for estimate for two brochures based on preliminary specifications.

Packaging: Shrink wrap in stacks of 50 with chipboard stiffener, pack 10 packs of 50 per carton, and stack on skids not to exceed 4 ft. in height.

Freight: Elements #1 and #2 to ship best way, one address in Chicago area, ZIP code 60015

ELEMENT #2—REFERENCE GUIDE 28-PG. + COVER BROCHURE

Quantities: 1,000   2,500   5,000   10,000

Cover: Flat size 7½ in. + bleed × 6 in. + bleed

28 pages: Flat size 7½×6 in., inside does not bleed

Finished size: 3¾×6 in.

Cover

Prints 4 color process + overall coating/1 PMS

Outside of cover is heavy coverage. Full-bleed image on front cover and solid screen mix on back.

Inside cover line art and type light coverage, no bleed

Paper: Lustro Gloss Cover 100#

Inside Pages

Prints 1 PMS, 2 sides

Inside pages are line art and copy light coverage, no bleed both sides.

Paper: 80# Patina Matte Book

Press checks: Customer press check for outside cover only, one-color forms to run to plant OK.

Bindery: (1) Cover score and fold; (2) collate, fold, and stitch inside pages into folded cover; and (3) trim book to final size.

Packaging: Bulk-pack 500 per carton, and stack on skids not to exceed 4 ft. in height.

Figure 1-2 (continued). A sample request for estimate for two brochures based on preliminary specifications.

Freight: Elements #1 and #2 to ship best way, one address in Chicago area, ZIP code 60015

Please furnish the following time requirements:
From receipt of disk to proofs _____ days
From approved proofs to press _____ days
From press to bindery _____ days
Ready to ship _____ days

Timeline from Presentation of Layouts
Both Brochures

Step	Working days	Running total
From approved concept		
Presentation to client of all elements	Start of project timeline	
Concept direction to creative	4 days	
Revised concept presentation	5 days	
Image concept and style approval	2 days	
Layouts to legal department for approval	10 days	
Release images for scan and color correction		
First round loose color and low-res FPO images on disk	3 days	
Final round of loose color approved	1 day	
Build mechanicals with FPO low-res	1–2 days	
Approved composite mechanicals	1 day	
Release disks and marked-up mechanical printouts to vendor		
Get press imposition proofs and constructed lasers	3 days	
Proofs approved	1 day	
Plate, print, and finish	5 days	
Packaging	3 days	
Ship	@ 8 weeks after concept approval	

Figure 1-2 (continued). A sample request for estimate for two brochures based on preliminary specifications.

PRELIMINARY BUDGET AND TIMELINE
FOR COLLATERAL MATERIAL

Company name _____
Date _____ Date due _____
Contact person _____

Project name _____
Number of elements in project _____
Element description _____

Finished size _____ Flat size _____
Image area _____

From receipt of disk with final files and approved loose color:
Composite color proofs
Digital _____ Film generated _____ On stock _____

Backed-up Lasers
Flat _____ Folded/constructed _____

Estimated time from receipt of files to proofs: _____ to _____ working days

Quantities to estimate: _____ _____ _____

Number of pages: _____ + Fly _____ + Cover _____

Body stock (# pp.) _____

Print information (complete information for front and back of forms)
Front: 4-color process _____ PMS _____ Match color _____
       Coating _____ Bleed _____ Coverage _____

Back: 4-color process _____ PMS _____ Match color _____
      Coating _____ Bleed _____ Coverage _____

Body stock (# pp.) _____

Print information (complete information for front and back of forms)
Front: 4-color process _____ PMS _____ Match color _____
       Coating _____ Bleed _____ Coverage _____

Back: 4-color process _____ PMS _____ Match color _____
      Coating _____ Bleed _____ Coverage _____

Figure 1-3. A preliminary budget and timeline for collateral material.

Fly sheet (# pp.) _____

Print information (complete information for front and back of forms)
Front: 4-color process _____ PMS _____ Match color _____
       Coating _____ Bleed _____ Coverage _____

Back: 4-color process _____ PMS _____ Match color _____
       Coating _____ Bleed _____ Coverage _____

Cover stock (# pgs.) _____

Print information (complete information for front and back of forms)
Front: 4-color process _____ PMS _____ Match color _____
       Coating _____ Bleed _____ Coverage _____

Back: 4-color process _____ PMS _____ Match color _____
       Coating _____ Bleed _____ Coverage _____

Special instructions:
Diecutting _____   Score _____
Embossing _____   Foil stamp _____
Foil emboss _____   Mounting _____
Film lamination _____   Tape _____
Gluing and folding _____   Other _____

Binding method _____

Packaging
Number of pieces per package _____
Shrink-wrap _____ Box _____ Kit _____

Does this element pack with other elements? _____
If so, which ones? _____

Estimated cost for printing, finishing, and packaging by quantity:
Quantity _____ $_____
Quantity _____ $_____
Quantity _____ $_____

Estimated time from mechanicals to finished materials ready to ship:
____ to ____ working days

Figure 1-3 (continued). A preliminary budget and timeline for collateral material.

POP AND POS MATERIALS
REQUEST FOR PRELIMINARY BUDGET
AND TIMELINE

Company name _____

Date _____ Date due _____

Contact person _____

Project name _____

Number of elements in project _____

Element description _____

Finished size _____ Flat size _____

Image area _____

From receipt of disk with final files and approved loose color:

Composite color proofs

1 side _____ 2 side _____ Quarter size _____ Half size _____ Full size _____

Die line ☐ Yes   ☐ No

Estimated time from approved high-res mechanical files to composite proofs:
_____ to _____ working days

In-Position Lasers

1 side _____ 2 side _____

Flat _____ Folded/constructed _____

Estimated time from approved composite film to see bluelines:
_____ to _____ working days

Quantities to estimate: _____   _____   _____

Base stock _____

Print information (complete information for front and back of forms)

Front: 4-color process _____   PMS _____   Match color _____

         Coating _____   Bleed _____   Coverage _____

Back: 4-color process _____   PMS _____   Match color _____

         Coating _____   Bleed _____   Coverage _____

Figure 1-4. A sample request for preliminary budget and timeline for POP and POS materials.

List PMS numbers _____ _____

_____ _____

_____ _____

_____ _____

Special instructions: Please indicate in line or offline

Diecutting _____ Score _____

Embossing _____ Foil stamp _____

Foil emboss _____ Mounting _____

Film lamination _____ Tape _____

Gluing and folding _____ Other _____

Packaging

Number of pieces per package _____

Shrink-wrap _____ Box _____ Kit _____

Does this element pack with other elements? _____

If so, which ones? _____

Estimated cost for printing, finishing, and packaging by quantity:

Quantity _____ $_____

Quantity _____ $_____

Quantity _____ $_____

Estimated time from mechanicals to finished materials ready to ship:

_____ to _____ working days

ATTACH MARKED-UP HARD COPY TO REQUEST.

Figure 1-4 (continued). A sample request for preliminary budget and timeline for POP and POS materials.

LOOSE COLOR REQUEST FOR QUOTE

Company name _____

Date _____ Date due _____

Contact person _____

Phone number _____ Ext. _____

Email address _____

Alternate contact_____

Phone number _____ Ext. _____

Project reference _____

Total number of images _____

Number of images on disk _____

Number of images to scan _____

Transparencies/digital images

Number	Source	Pathed	Scan sizes	DPI	Retouch	System work
_____	_____	_____	_____	___	_____	_____
_____	_____	_____	_____	___	_____	_____
_____	_____	_____	_____	___	_____	_____
_____	_____	_____	_____	___	_____	_____

Marked-up hard copy prints attached: ☐ Yes ☐ No

Special instructions:

_____

_____

Illustrations boards/disk

Number	Image size	Scan sizes	Pathed	DPI	Retouch	System work
_____	_____	_____	_____	___	_____	_____
_____	_____	_____	_____	___	_____	_____
_____	_____	_____	_____	___	_____	_____
_____	_____	_____	_____	___	_____	_____

Marked-up hard copy prints attached: ☐ Yes ☐ No

Special instructions:

_____

_____

_____

Figure 1-5. A sample request for quote for loose color.

First round proofs:
Half size _____   Full size _____
Other _____   How many sets_____
Digital proofs _____   Film proofs _____

Second round proofs:
Half size _____   Full size _____
Other _____   How many sets_____
Digital proofs _____   Film proofs _____

Low-res images on disk: ☐ Yes   ☐ No

Final images on disk: ☐ Yes   ☐ No

Place final image files in mechanical: ☐ Yes   ☐ No

Estimated time from receipt of images to first-round color: _____ to _____ working days

Estimated time from receipt of revisions and corrections to second-round color: _____ to _____ working days

Estimated cost for scans, color corrections, system work, and loose proofs: $_____ to $_____

Estimated cost for each additional set of proofs: $_____ to $_____

Estimated cost for each additional round of color work and proofs: $_____ to $_____

Estimated cost for each additional hour of system work: $_____

Figure 1-5 (continued). A sample request for quote for loose color.

2
Sizes and Configurations

Sizes and configurations include the dimensions of the piece and the folding and/or finishing required. The terms and order used to state this information are important. If not given properly, there can be misunderstandings that can affect the estimate. To communicate this information properly, there are only a few basic rules.

Dimensions

The rule on stating dimensions is one that very often is not followed, and it can cause real problems. When stating the size of any two-dimensional printed piece, whether it is a brochure, a sell sheet, or even an envelope, always state the size as *width × height.* Here are some examples that will help to explain the effect of giving dimensions in the wrong order.

Example 1. A four-page, 8½×11-in. newsletter with no bleed would be stated as a flat size 17×11-in. sheet that is folded to 8½×11 (**Figure 2-1A**).

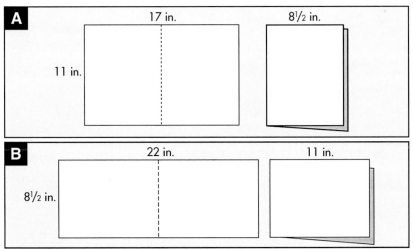

Figure 2-1. (A) A four-page 8½×11-in. newsletter with no bleed. **(B)** A four-page, 11×8½-in. newsletter with no bleed.

If stated as an 11×8½-in. newsletter with no bleed, the flat size would be a 22×8½-in. sheet, folded to 11×8½ in. (**Figure 2-1B**). As you can see, while the piece has the same finished measurements, the visual impact and the flat size are very different. If an estimate were being generated from these stated dimensions, the amount of paper required, the number of finished pieces from a parent or full sheet of the paper, and the press size needed to print it would all be wrong for one of these examples if they were stated for the other.

Example 2. A "shelf talker" with a printed area of 9×4 in. with no bleed and a 3-in. fold-back has a flat size of 9×7 in. (**Figure 2-2A**). If the size were given as 4×9 in. plus a 3-in. fold-back, the flat size would be 4×12 in. (**Figure 2-2B**).

In addition to the size issue, this element would require a score for the fold-back. Consequently, the grain direction of the paper could be a factor. If so, the grain direction of the parent-size sheet would need to be considered. This could affect the parent-size sheet being used, could increase the waste factor, and could result in higher paper costs.

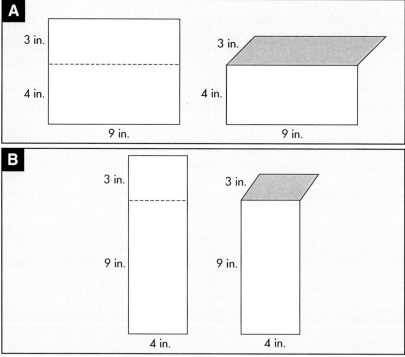

Figure 2-2. (A) A "shelf talker" with a printed area of 9×4 in. with no bleed and a 3-in. fold-back. **(B)** A "shelf talker" with a printed area of 4×9 in. with no bleed and a 3-in. fold-back.

Example 3. If a "header card" has an image area of 26×30 in. with an 8-in. foot and ¼-in. bleed on all four sides, the print area would be 26½×38½ in. (**Figure 2-3A**), a size that could be printed on a 40-in. press. If the header card had an image area of 30×26 in. with an 8-in. foot and ¼-in. bleed on all four sides, the print area would be 30½×34½ in. (**Figure 2-3B**). This image size exceeds the maximum image size of a 40-in. press and would require an oversize press of 54 in. or larger.

The 40-in. press, a much more common configuration than the larger-format presses, can accommodate a sheet 28×40 in. The print area is approximately 27⁹⁄₁₆×39½ in. including bleed (the print area varies slightly from press to press). The press sheet sizes that can be run on 40-in. presses are available in virtually all paper stocks, while the oversize materials are available in only some stocks or must be special ordered.

When a piece has a number of folds or diecuts, it is sometimes necessary to furnish a number of dimensions to convey the size of a piece, the flat or overall size, the image area, and the finished size. It is always better to err on the side of caution. The more information you give on any job, the better the chances there will be no misunderstanding regarding the format, at the finished piece will look like, and so on. If the folding or finishing of a design is really complicated, drawings and composites (comps) can help communicate the request or define the estimate.

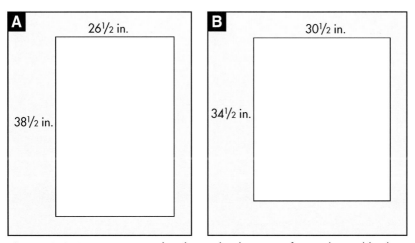

Figure 2-3. (A) A 26×30-in. header card with an 8-in. foot and ¼-in. bleed on all four sides has a print area of 26½×38½ in. **(B)** A 30×26-in. header card with an 8-in. foot and ¼-in. bleed on all four sides has a print area of 30½×34½ in.

Page Count

When determining the number of pages or panels in a brochure, catalog, or any printed multi-page publication it is a matter of simple multiplication and division.

Folded Material

Folding a single sheet of paper in half produces a 4-page signature (**Figure 2-4**). This is the basis for counting pages in all single-fold material. Working from the front open edge, the outside is page 1, the inside page to the left of the fold is page 2, the inside page to the right of the fold is page 3, and the back outside is page 4. Therefore if the piece you are working on has four flat sheets of paper folded in half to form a book, it is a 16-page book: 4 sheets of 4 pages (4 × 4 = 16).

Conversely if you have a 20-page book and want to determine how many flat sheets of stock it took to make it, divide 20 pages by 4 pages per sheet (20/4) to determine the answer: 5 sheets.

Books and magazines that run on web presses are usually printed as 16- or 32-page signatures. The number of pages in a signature is determined by the page size and the paper size. If you are producing a 5½×8½-in. book that will be printed on a web press, it will most likely be printed as a series of 32-page signatures. Therefore, to make printing as simple as possible, the total page count in the book should be an even multiple of 32 pages; e.g., 64, 96, 128, and so on. If the page count is not an even multiple of 32, you could include one 16-page signature. If your page count was 144, you would print four 32-page signatures and one 16-page signature. Always check with the printer

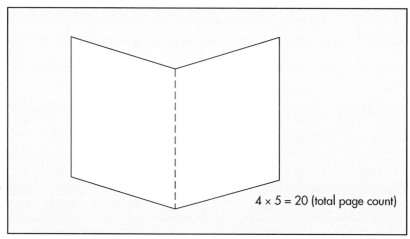

4 × 5 = 20 (total page count)

Figure 2-4. A 20-page booklet would consist of five flat sheets that have been folded once to form 4-page signatures.

and/or bindery to determine which signature sizes to use in your planning procedure.

This seems very elementary, but what if a designer is working in reader spreads or on individual page layouts and the copy flows to fill 14 pages. It might look right in the layouts, but when you attempt to build a comp the problem becomes immediately apparent.

So it is always a good idea to comp up materials to be sure you have the correct page count. When the count is off, it is not always a simple matter of adding a couple pages or taking out a page to fix the problem, because the addition or deletion can affect the design concept.

The exceptions to the "divisible by 4" rules on page count are pieces that have to be folded more than once to be at final size. Following are brief discussions of some of these exceptions.

Gatefolds, which are panels folded in from the outside edges of a piece (**Figure 2-5**), can be added to some designs to give you two more panels. In some cases the amount of copy and its relationship to the layout require that a specific block of information appear on a single spread and that block won't fit on a standard two-page spread. To accommodate the additional copy, you could increase the size of a two-page spread by adding a foldout panel to one or both sides of the spread.

Folding a flat sheet into three panels to creates a single foldout panel called a ***single gatefold.*** The sheet, which is wider than the standard sheet by the size of the foldout, is folded in two places as shown in **Figure 2-5**.

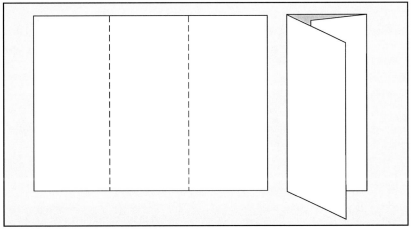

Figure 2-5. A single gatefold. The panel that is folded inside the other two panels should be at least ¼ in. narrower.

When bound into a set of 4-page sheets, it will increase the page count of the finished piece by 2 pages. If the base book was 20 pages, the addition of the gatefold would be indicated by specifying a "20-page plus single gatefold." When asking for a vendor quote, you would need to indicate the width of the gatefold and whether it is to the right or left and where it falls in the book.

A *double gatefold* (**Figure 2-6**) has two foldout panels. It starts as a flat sheet that is wider than the standard sheet by the width of the two foldouts. The sheet is folded in three places, as shown in **Figure 2-6,** to create a spread nearly twice the width of the book. You would have to indicate the position and width of the panels, and the job would be specified as a "20-page plus double gatefold."

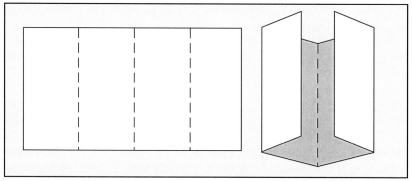

Figure 2-6. A double gatefold. The two outside panels should be at least ¼ in. narrower than the other two panels.

The added panel or panels will not be the full width of the finished piece, the panels must end short of the gutter on the cut edge so that it will not be caught in the fold and wrinkle. In addition, the folded edge must fall to the inside of the cut edge of the piece, so that the fold will not be cut off when the piece is trimmed to final size. The most accurate way to calculate the exact size of each panel is to make a blank comp in the weight and type of stock that will be used in the finished piece. After trimming it to the finished size, measure each of the panels to determine their size.

Where double gatefolds fall in the book determines how they are viewed. For instance a double gatefold on the center spread gives you foldouts to the left and right of the center spread.

The placement and folding direction can make single gatefolds work like double gatefolds without the placement limitations (**Figure 2-7**). Gatefolds can be a very practical way to increase print area without having to add another four pages or change the overall size of the piece.

Let's look at the possible impact on the press layout of a full-sheet single gatefold. If the finished size of the book is 8½×11 in., the 4-page spreads would have a flat size of 17×11 in. A single gatefold would

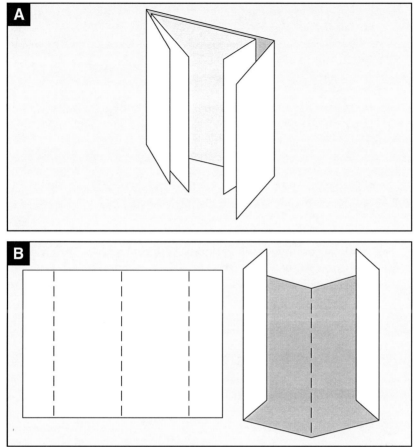

Figure 2-7. The use of single gatefolds to function like double gatefolds but without the placement limitations: **(A)** two 6-panels to form 8-panel and **(B)** an 8-panel with short panels.

have a flat size of approximately 25¼×11 in., because the panel can not fold all the way to the center or it would be creased on the edge by the fold. The 4-page spread could be printed on a 12×18-in. press, but looking at the dimensions of the gatefold clearly shows that the 12×18-in. press would not accommodate the gatefold. Therefore, the job would have to be quoted on a press that could accommodate at least a 26-in. press sheet. For an 8-page full-sheet double-gate piece, the press format required would be even larger.

Just to be sure this is clear, here are a few additional sample calculations for reference:

- An 8½×11-in. single gatefold with one 4-in. panel would have a flat size of 21×11 in.

- An 8½×11-in. double gatefold with two 4-in. panels would have a flat size of 25×11 in.
- An 8½×11-in. double gatefold full-sheet would have a flat size of approximately 33½×11 in.

Cut-Sheet Bound Materials

When counting pages for cut-sheet materials like GBC, Wire-O, spiral, or three-ring binders, count each side of the sheet as a page. For example, a stack of 16 sheets would be 32 pages (16 × 2 = 32).

It can be confusing when communicating page counts for these kinds of materials because you can have any number of pages divisible by 2, instead of 4. In addition, with cut-sheet bound materials, you will often not print the backs of the sheets. Therefore, it is a good idea to use both sheet count and page count when describing a cut-sheet bound book; e.g., a 16-sheet, 32-page book.

You can incorporate single gatefolds in cut-sheet bound material. However, instead of working with a maximum flat sheet size that is 1.5 times the width of a two-page spread, you are generally limited to a flat sheet size that is slightly less than twice the width of an individual page. The binding method used will determine how far in you can fold the panel without hitting the binding area or the rings of a three-ring binder.

None of the above considered the use of different weights or types of stock or paper in a book or brochure. Since this affects how page count is reported in a specification, let's briefly discuss these issues.

Self-Cover and Plus Cover

Generally a book is a self-cover if all the pages in the book are the same weight and paper. You can have a self-cover book with a gatefold inside the last page, and it would still be a self-cover. A lot of people, however, make the mistake of referring to it as the inside back cover, which can be confusing if the specifications do not include a cover stock.

When you are working on a book that has a different cover stock than the inside pages, whether it is simply a different weight or a completely different stock (e.g., text cover and coated book), count the book and cover separately. Also, state the inside page count and then the cover. This is often just stated as "plus cover" in the specifications.

Here are some examples of how to state page count specifications:

- "20-page self-cover with single gatefold on the inside of the last page" describes a book with five sheets of paper, four of which are folded once and one sheet that has two folds.
- "16-page plus cover with single gatefold on inside back cover" describes a book that has a 4-page cover with a single gatefold wrapped around four sheets of paper folded once.

Fly Sheet

A *fly sheet* is a transition sheet that falls between the cover and the body of a book. Fly sheets are usually a different color or texture than the cover or inside pages. Very often they are lightweight translucent paper. When stating the sheet count for a book with a fly sheet, it is separate from the body and cover page counts. Again, this is to identify the fact that a third type of stock will be needed to produce this book. The description would be "16-page + 4-page fly + cover."

Multiple Signatures in a Saddle-Stitched Book

When working on an annual report or brochure that is to be folded and saddle-stitched where the design calls for a differentiation from one section of the book to another, a different stock can be used for each section, but usually no more than two. Each type of stock is folded separately and then bound into the cover independently of one another. The specifications would be stated as "16 pages gloss + 16 pages uncoated + 4-page text cover, 4 wires on the spine." This indicates to the vendor that there are three distinct paper stocks required in this book.

Pockets and Flaps

Pockets or flaps can be horizontal, across the bottom edge of the piece, or vertical with folds parallel to the spine (**Figure 2-8**). Pockets are usually on the inside front or back cover or both, and there can be different configurations on each. Pockets and flaps can be diecut in any shape or cut square. The height or width can be as deep as you want as long as it does not reach the center fold or exceed the edges of the folded piece. You are limited only by the size of the parent sheet of the stock you have chosen and the budget. The larger and more complicated the configuration, the more expensive the die to cut it.

The difference between a flap and a pocket is the finishing. Flaps are simply panels that are folded in (unsecured), while pockets have tabs that secure the panel to the body of the piece. The configurations and function determine whether the pocket flap is secured by one tab or two.

Blank Comps and Dummies

You can request from your vendors or paper merchant a blank construction of any project, be it brochures, pocket folders, invitations, or envelopes, to name a few. No matter what you are going to be producing, always insist on the creation of a **blank comp** or **dummy.** The blank comp helps to ensure that the stock weight is correct, the overall size is proper, and provides the opportunity to feel and see the actual piece.

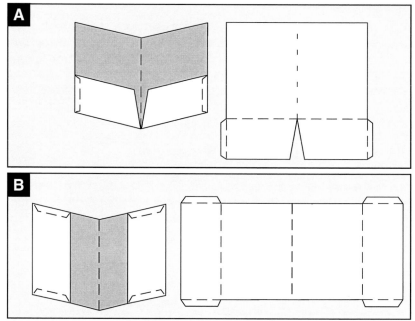

Figure 2-8. (A) A pocket folder with horizontal pockets. **(B)** A pocket folder with vertical pockets.

Paper vendors can also furnish printed samples of the paper stock or stocks being considered for your project, as well as blank flat sheets for ink drawdowns if the project is being printed conventionally.

Blank comps are very useful in determining layout and live area. You can also use them to show clients how the final piece will look and feel, to ensure that the piece has the weight, feel, color, and composition that the client is expecting. Blank comps, which are usually furnished without charge, are a good way to avoid unpleasant surprises and make the production process smoother.

Most paper merchants have a staff trained to create blank comps to your specifications. If you are not sure how to request the materials you need, call your vendor and ask that they have dummies or mockups made for you in the stock and weights your are considering.

3
Photography and Illustrations

The field of photography is expanding and changing due to the advances in electronics, but the basics remain the same. A photograph is an image captured through a lens in one dimension, either black-and-white or full color. There are a number of sources for photography, and they are discussed here briefly.

Commercial Photography

When a piece requires a photographic image, there are a number of ways to proceed. The way to produce an image that most closely resembles the concept would be to hire a photographer and set up a *photo shoot.* Using a copy of the approved layout, the commercial photographer will set up the shoot and light it to match the layout as closely as possible so that the photo will need little or no system work.

Doing a photo shoot can require the services of a number of people in addition to the photographer and an assistant. Sometimes a stylist is needed. This person handles the arrangement of the elements of the picture. The stylist is usually specialist in a specific area, like food or fashion, and it is the stylist's job to prop, accent, and blend the elements to get the best visual layout. In addition, the stylists will handle arrangements for the models, make-up artists, hairdresser, and wardrobe people necessary for a successful photo shoot.

The costs involved in a photo shoot are determined by several factors: e.g., how many images are being shot; the number and types of people required and their fees; the costs of any props or rentals for the visual elements; location rentals; and costs for the photographer's time, equipment, and usage fees; and sometimes—although it is rare—there are costs for film and processing.

Images are reviewed, and the best image or images selected, then the prepress processes can begin. If the shoot was done on film, image(s) will need to be scanned to digital format.

Once the images are in digital format (scans or Raw files from the photographer), there are a number of things that will need to be done such as sizing, cropping, retouching, and color correction. Working from an approved layout, the image can be cropped, then proofed and reviewed with the creative director. Marked-up proofs with direction on color, retouching, cleanups, (direction to remove any stray highlight or odd visuals) are used by the vendor to ensure that they make all of the necessary changes and correct and balance the color as requested.

Digital Photography

The improvements and advances in this technology have been remarkable and have made digital photography the current standard for most commercial photographers.

Figure 4-1. High-resolution digital camera backfitted to conventional large-format camera. (Courtesy Phase One)

With these cameras, a ***charge-coupled device*** (CCD) captures the image that shines through the lens as digital information and records this "digital image" directly to a disk or computer hard drive with no film or processing required. Some digital cameras have a preview option that lets you view the image prior to actually capturing the image. This allows you the choice of saving an image or not once it appears on the screen. Digital cameras have become very portable and economical. Images can be previewed both on site and remotely via digital files on a computer screen. Large-capacity memory chips and instant-download capabilities make the number of images that can be taken and stored virtually unlimited.

Since there will be no transparency to scan, it is important to know the largest final size of the image to be used and at what resolution in order to be sure that the digital image is the correct size and resolution for the finished piece.

Digital photography can save time because there is no delay for processing and there is no need to do a scan since the image is in digital format. Digital photography, however, does not eliminate the need for prepress processes—color correction, retouching, and proofing—because photographic images are basically the same regardless of the source of the image.

Stock Photography

Sometimes a generic image that fits the concept can be found from the catalogs or websites of companies that handle images on virtually every subject and sell their rights to them. The images from these companies are called ***stock photography.***

Generally the fee for the use of these images is determined by where it will appear, how many copies of the piece will be printed, and the duration of the program. If the image is to be used in wide distribution or on multiple elements or over a long period of time, the fees for the use of a stock photograph can be more costly than working with a photographer to do a photo shoot with a negotiated image buyout.

To order a stock image, the first step is to select a specific image. Most photo libraries can be searched online for specific type (i.e., image type, royalty free, rights managed, etc.) and keywords to narrow down the selections. Once the image(s) have been selected, you can download low-resolution images for use in layouts.

Royalty-free images are sold for a flat rate, and there are no restrictions on the usage. Conversely, you cannot ensure exclusivity for these images. Some royalty-free images are sold as part of a collection of images, and you must purchase them as a set. These can often be ordered on a CD or DVD.

Rights-managed images' fees are determined by the specific usage, medium, and length of time, country of reproduction, and the size and resolution of the file needed. You can also ask for the history of usage on specific images. This allows you to avoid things like conflict in category. It is also sometimes possible to purchase exclusive rights to the image in a geographic area, business category, or both. Once all the details have been defined, the usage fee can be negotiated and the image purchased. The high-res image is generally posted by the service and a link sent so that the image can be downloaded directly to your hard drive.

Online stock photography sources offer digital images that can be ordered online and downloaded directly. The fees for these images are usually based on the file size requested. Before ordering a digital image, it is important to confirm that the image size is large enough and that they have enough resolution for the project.

Stock images are available from a wide variety of specialty sources. Some publishers archive images that they own the rights to and, for a fee, will search their files and send you images for consideration. Images are available from professional sports organizations. They usually require that the client have a license to use their material, and there are very specific rules about the use of their images as well as image fees. In some cases, fees have to be paid to the individual player or players, and specific credit lines are required when using photos from these sources. Images from these sources are generally on transparencies and will need to be scanned and returned.

There are digital image services that sell discs of live images by subject, like foods. These discs carry no usage fees past the purchase price of the disk. There are also royalty-free image disks available. If the images are large enough and have enough resolution, these options can reduce the cost of images for the piece. Be sure to check the size and resolution of these images before finalizing the budget. If they are too small or have limited resolution, the high-resolution image will have to be secured or shot.

Furnished Images

A client may also own the images that are to be used. These are referred to as *furnished images.* These images may have been used in a previous printed piece and may be in digital form or exist only as a transparency. The first thing to check when a client furnishes a digital file is the resolution and size of the image to be sure that it is large enough for the project. It is better to address such issues when the prepress estimates are being done.

Illustrators and Illustrations

Illustrations encompass original art done in any number of styles and mediums. Techniques range from very primitive styles to cartoon art to photo illustration. The layout and feel of the piece dictates the choice of illustrator.

Illustrators often specialize in a specific era or subject matter, like fantasy art, '50s stylized drawings, people, or cars. Some work in only a particular medium (e.g., paints, chalks, or pencils) or on a particular type of material. Some illustrators have adapted their styles to combine hand drawings with the computer. They will sketch out an image and then scan it into the computer where it is completed. The style, medium, source, and whether a computer is used are all factors considered by the art director and/or creative director when choosing an illustrator for a specific project.

If the final illustration is done on a computer, it is a digital file. The digital layout can be processed through the proofing stage in prepress to check the final color, since not all computer monitors are color-correct. In some cases, system work to marry the illustration to another image or a product shot will be needed.

In order to write the preliminary prepress specifications and set a timeline, it is necessary to identify the source of the images and determine what needs to be done with them before they are ready for release in the final layout file or on film. Knowing the lead time required to secure the photography or illustrations is also necessary to determine the release dates for a production timeline.

4
Paper

Printing can be done on plastic, metal, or cloth, but paper is by far the most common material used in the production of a printed piece. Paper represents a wide and varied range of options in quality, type, finish, weight, and price. Papers are available for just about any visual effect you can imagine, in a rainbow of colors, and in a variety of textures, finishes, coatings, weights, and grades. The choice of paper stock is determined by a number of factors, but the most important considerations are the quality, color, surface, opacity, weight, availability, and cost—but not necessarily in that order.

This chapter discusses the characteristics of paper and surface options of the various types of paper, as well as their recommended uses. Having so many options in paper can be daunting, but by using the basics in this section and working with paper merchants and samples, you can make advised decisions.

General Information

The base ingredient of paper is *pulp,* which is the fibrous material for papermaking produced either mechanically or chemically from cellulose raw material. The primary sources of these fibers are wood or cotton. Pulp is treated with chemical additives to brighten, tint, and add opacity to the material. Whether the fibers come from virgin sources (e.g., trees, cotton plants, or other organic growth materials) or are recycled from previously manufactured goods (e.g., used paper products or old newspapers), the material must be reduced to fibers, cleaned, and worked into a state that allows cohesiveness in the papermaking process. The quality and characteristics of the paper are affected by the source of the fibers, as well as how the pulp is processed.

Due to the ever-increasing demand for paper products, sources estimate that the worldwide demand grows at a rate in excess of 3% per year. It has become imperative that alternate or nontraditional sources of fiber be developed. Some of these sources are organic and have spawned lines of *"wood-free" papers.*

In some cases, nontraditional pulps have been used in products that previously used high-absorbency wood pulp. These products, called *fluff pulp,* are soft and absorbent materials that can be used predominantly in nonprinting products like disposable diapers and hospital pads, and they free up a large amount of traditional pulps.

Recycled Paper Products

Recycled materials are reclaimed from manufactured goods and used to make new products. The source of the recycled material affects the amount of processing required to reclaim the fibers as well as the use of those fibers.

Internal by-product is the term used for the reclaiming of fibers lost in the papermaking process, much like the fiber recovered from the water removed from the papermaking furnish on the paper machine. This category also includes the waste from the rewinding and trimming processes as well as materials that are reprocessed because they did not pass the quality control inspection. These materials have never been used for another purpose and, therefore, have no outside contaminants to be removed.

Preconsumer waste is the plain and printed materials that have not been used by an end user—e.g., undistributed or obsolete forms, manuals, or letterheads; magazines and book overruns that are complete but never leave the warehouse; as well as waste materials like scraps from envelope converters and notebook manufacturers. These are all considered preconsumer waste. This fiber can be used in a higher concentration because the de-inking process yields cleaner fibers.

Postconsumer waste is any fibrous material that is found in companies, homes, offices, schools, municipal landfills, and so forth. These materials are not all suitable for recycling into printing papers so they have to be sorted by the recyclers before they can be processed. Basically anything that has been used in the marketplace and then recovered for reuse is considered postconsumer waste.

The type of paper being manufactured and the source of the recycled material determine the use of the recycled fibers and the concentration of those fibers in a particular product. In many cases you will find that the mill indicates the amount and source of the recycled material in a line of paper.

In addition to the obvious reduction in the number of trees harvested for the papermaking process, recycling of paper materials helps to reduce landfill materials by a significant amount. The recycling process of collecting, sorting, transporting, de-inking, and repulping make the cost of recycling as high as or higher than the harvesting and converting of raw materials.

It has been said that for every tree that is saved by recycling, another natural resource is reduced. While this statement is valid because of the fuel consumption in the transportation of the materials to the mills and the energy required to reclaim the fibers, we must consider that for every ton of recycled material, there is one less ton of material in the landfills. In the early 1990s, it was estimated that up to 40% of the material in landfills was paper.

Wood Pulp

At this time, the primary source of pulp is still trees. Logs from trees can be processed in a couple of different ways, each designed to yield a specific type of pulp.

Debarking (**Figure 4-1**) is the first step in converting cut timber into pulp fibers. The logs are fed into a large rotating drum with sharp extrusions inside. These extrusions are designed to remove the bark from the logs as they tumble inside the drum. Once the logs have been debarked, the type of pulp being made determines the next steps.

Free sheets are paper made from ***chemical wood pulp.*** This process generates pulp for fine printing paper. When this kind of pulp is made, the resin, knots, and impurities are removed from the base material, and the material is reduced to individual fibers.

After the debarked logs are chipped, the chips are cooked under pressure in a series of digesters using chemicals to release the impurities.

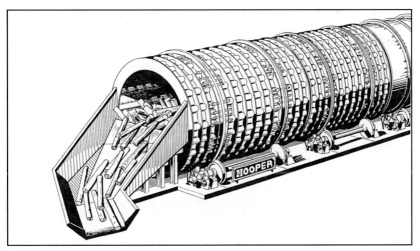

Figure 4-1. A barking drum. The logs enter the drum; are debarked by impact, compression, and shear forces; and are continuously discharged. (Courtesy S. W. Hooper Co.)

The heat and pressure allow the fibers to separate gently. The digesters are like a series of processing units that draw off the impurities and resins, washing the fibers clean. Draining the residue out of the materials washes away the impurities.

The characteristics of the fibers vary depending on the type of tree. For instance, in soft wood trees like the southern pine, the fibers are longer and thicker than the fibers from hardwood trees. The length and thickness of the fibers affects the surface characteristics and density of paper. The specific sheet characteristics desired determine the ratio of combined fibers from various sources necessary to achieve the right mix of fibers.

The chemical pulping process is time-consuming, and the yield of pulp from the raw materials is approximately half of each log. The time and reduced yield are factors in the cost of the papers made from this pulp. The paper made from chemical wood pulp is usually consistent to the specifications of the grade in brightness, color, gloss, and formation.

A less-expensive process used to reduce the debarked logs to fibers is a high-yield, low-quality process called **groundwood** or **mechanical pulp.** This process utilizes nearly the entire log. The debarked log is pressed into a grinding stone, and the resulting material resembles sawdust. The very nature of this process makes no allowances for the removal of the impurities or irregularities from the logs. The fibers yielded by this process will have a high retention of impurities and transient elements, not all of which will be removed in the process of making pulp. Therefore, the paper made from the fibers will be less stable and will more readily discolor.

The grinding process (**Figure 4-2**) cuts the log's natural fibers into short fluffy fibers. The material is cleaned, bleached, and processed to create the flow of cellulose in a much shorter time frame. Since the

Figure 4-2. A two-pocket grinder for manufacturing groundwood pulp. Debarked logs are pressed against a revolving stone in the presence of water. (Courtesy Montagne Machine Co.)

pulp is not as refined, paper made from this pulp will not be as consistent, the surface will not be as smooth, and the formation is not as regular. Nevertheless, mechanical pulp is not without uses and advantages. The combination of the high yield from each log, reduced processing costs, and high opacity-to-weight ratio makes it possible to use a lighter weight of stock with less show-through. Keep in mind the requirements of the stocks for which pure groundwood pulp is generally used. For instance, the primary requirements for newsprint are that it be an economical, lightweight paper with sufficient opacity to allow printing on both sides with minimal show-through. These requirements are all met by groundwood pulp.

A wide range of papers are made using a mix of groundwood and chemical pulp. The combination of these materials allows the mills to make papers that utilize the best properties of each kind of pulp. The percentage of groundwood to chemical pulp varies depending on the characteristics of the sheet the mill is making.

The amount of groundwood pulp added to the base stock of a sheet influences the opacity, the surface, and the bulk and is a factor in determining the grade of the finished sheet. Even a small percentage of groundwood pulp can increase the bulk and opacity of a sheet over a similar weight and grade that does not have the groundwood content, but the surface and color can be affected slightly in adverse ways. The higher the percentage of groundwood pulp in a sheet, the lower the grade of the paper. The improved bulk-to-weight ratio and opacity often offset the effects on the quality of the surface, color, and stability since these papers are used primarily for long-run brochures, catalogs, and publications that do not need to have extremely long shelf lives. Since paper costs are calculated by weight, another advantage to groundwood content sheets is that the increased bulk and opacity can make it possible to use a lighter sheet; a lower basis weight for a piece can mean a savings in paper costs, freight, and postage on large orders.

Making Paper from the Pulp

The pulp is put into a beater with a large quantity of water and treated prior to being sent to the papermaking machine. For each grade or quality of paper made by a mill, there is a formula that dictates the final characteristics of the sheet. The formula of a sheet includes the type and amounts of the different pulps, additives, binders, brighteners, pigments, and so forth.

The *beating* process prepares the pulp for the papermaking process by fraying the cellulose fibers. The duration and intensity of the beating process impacts the bulk and surface characteristics of the finished product. With a gentler and shorter beating, the fibers will be less cohesive. As a consequence, the surface of the sheet will be rougher, but the bulk or thickness of the sheet will be higher because the fibers themselves will not be as refined. Conversely, the longer and firmer the beating, the smoother and more cohesive the finished sheet and the lower the bulk. There are inevitable compromises in the papermaking process.

Once the beating process has been completed, the slurry consisting of approximately 99% water and 1% pulp is pumped into the **head box** at the top of the papermaking machine (**Figure 4-3**). The headbox keeps the fibers dispersed and delivers them to the forming section at the proper speed relative to the moving wire. Next, the fiber suspension enters the *slice,* an adjustable rectangular orifice that delivers across the wire width a continuous sheet of water-suspended fibers. As the fibers flow through the slice and increase in speed, they tend to align themselves in the direction of their flow. This pronounced fiber alignment accounts for the grain of paper.

A *fourdrinier* paper machine uses a horizontal wire belt as its forming section; the wire consists of finely woven bronze or plastic mesh. Fibers and water flow onto the wire with the water draining through the wire because of gravity. The fibers interlace in a random fashion to form a mat. As the wire moves forward, more fibers are deposited over the first formed layer, thereby building up a succession of fiber layers. The side of the fiber mat formed in contact with the wire is called the *wire side* of paper. The top side, or the side not in contact with the wire, is designated as the *felt side.* The felt side has more short fibers, fines, and filler than the wire side.

As the wire continues forward, it passes over vacuum boxes and rolls with perforated shells that suck water from the newly formed web. Next the paper web enters the press section, which removes even more water from the paper by pressing and suction. The press section also compacts the paper web and brings fibers into closer contact for better fiber bonding and sheet strength. It smooths the paper and has an important influence on the paper's final bulk and finish. High-bulk, antique-finish papers are given little wet pressing, whereas lower-bulk, high-finish papers need a greater degree of wet pressing.

The paper web then enters the drying section where a number of hollow, steam-heated cylinders remove even more moisture from the paper. As it dries, the web is kept under tension to prevent cockling, distortion, and uncontrolled shrinkage.

Paper Properties and Characteristics

Certain measurable characteristics of paper have been determined by the papermaking formula and the ingredients in the paper. Each of the following is either an additive or a property of the paper that

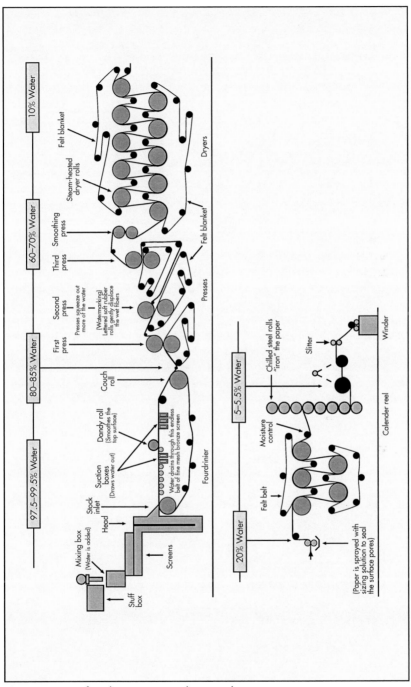

Figure 4-3. A fourdrinier papermaking machine.

impacts the visual presentation of the paper as well as the way the sheet will print.

Loading and fillers. The major nonfibrous raw materials fall under the term *loading,* or *fillers.* Loading is made up of finely divided, relatively insoluble inorganic materials or minerals—most commonly clay, titanium dioxide, and calcium carbonate—that are incorporated into the papermaking composition prior to sheet formation. These materials modify such characteristics of the finished paper as opacity, brightness, printability, texture, and weight. Loading is used to impart softness, reduce bulk, increase smoothness, make paper more uniformly receptive to printing inks, and lend greater dimensional stability. The prime reason for loading in printing papers is to increase opacity and brightness, reduce ink strike-through, and decrease the harshness of the fibers.

Formation. Formation indicates the structure of paper and the uniformity with which its fibers are interwoven and distributed. Formation is significant because of its influence on other paper properties. The levelness and smoothness of a paper, for example, are greatly dependent on the uniformity of its formation. The formation in an uncoated sheet will be much more apparent when printed than in a coated sheet, but it is important to consider in both. Formation is usually judged by viewing paper by transmitted light. The transmitted light will show dark and light spots, depending on the wildness of the formation.

Color. The hue or shade of the stock, whether the sheet is white or a color, can vary from run to run, as well as from weight to weight in the same stock. Mills create a unique formula of dyes and pigments for each paper and establish acceptable parameters of color variance for their sheets. The quality control technicians at a mill use these parameters to determine if there is a "shade shift" or color variance and whether it falls within the acceptable standards. If it does, the material can be shipped; if not, the material will be rejected and reprocessed.

Mills can only do so much to control the color shift from run to run due to a number of variables in the papermaking process, ranging from the source of the pulp and its color and configuration to variance in dye lots and pigments. In addition to the shift in color from run to run, there can be shade variations caused by the density of a sheet; a book and cover weight of the same stock, for example, may not exactly match.

It is always advisable to ask for stock samples. The approved samples can be compared to the run stock. If there is a big variation in color, it can be addressed before the printing process has started.

Color can be an issue even in whites. One white paper, for example, may have a blue cast and another a yellow cast. When color is a critical issue, compare stocks of the same grade to be sure that the paper chosen is the "white" best suited to the images being printed. When using multiple weights of a stock or mixing stocks, inspecting samples of the run stock is the only way to confirm that the stocks are compatible whites or colors.

Whiteness and brightness. Paper *brightness* may be defined as the relative amount of light reflected from the surface. Any color of paper can be bright—for example, bright red, bright green, or bright blue. To be white, paper must reflect all wavelengths of light in the visible spectrum at a high level. Papers that reflect the total visible spectrum are also known as "neutral" whites. Most printing papers, however, are anything but neutral. They come in many shades of "white." "Blue" or "cold" white describes papers that reflect a relatively higher percentage of light in the blue and violet region. "Warm" whites are papers have higher light reflectance in the red or orange region of the spectrum. Gray papers, on the other hand, reflect all wavelengths more or less equally (depending on their shade) but at a low reflectance, or brightness, level.

For measuring purposes, the paper industry defines brightness as the percent reflectance of blue light from a paper's surface measured at a specific wavelength—457 nanometers. Reflectance measurement at this wavelength is very sensitive to the detection of blue and yellow tints, as is the human eye. Most white papers are in the 60–90 brightness range, depending on type and grade.

How bright a paper should be depends on printing requirements, cost, and end use. The desire for a high printing contrast that commands the reader's attention may justify the added cost of high-brightness papers. Low-brightness papers are preferred for applications such as books that require minimum eye strain and ease in reading.

Opacity. The measure of the amount of show-through from one side of the paper to the other is called *opacity.* The less show-through, the better the opacity. Opacity in many sheets improves with the increase of weight and bulk, but there are specialty papers that are very thin and very opaque. These are sometimes referred to as ***Bible opaque,*** since they are used to print Bibles.

The design, how much if any show-through you can tolerate, amount of ink coverage, and the color of the ink usually determine opacity requirements for a specific job. Paper opacity is rated on a scale of 1 to 100, with 1 being transparent and 100 completely opaque.

Ink holdout. The extent to which paper retards the inward penetration of a freshly printed ink film is referred to as **ink holdout.** The better the holdout, the smoother and sharper the color will appear. Ink holdout is a consideration on all papers, to one degree or another, but it is easiest to see on coated sheets. The smoother and tighter the surface of the sheet, the better the ink holdout will be. When a sheet has good holdout, the ink will lie smoothly on the surface of the sheet.

On an uncoated or loosely coated sheet, the ink can be absorbed unevenly into the surface of the sheet; this is called **strike-through** or **strike-in.** The visual effect is an uneven ink density and dull flat color.

Grain direction. Paper grain, a function of fiber orientation and drying stresses, runs in the direction that paper travels through the paper machine. Papermakers refer to fiber orientation as being either in the **machine direction** (grain direction) or **cross-machine direction.** Printers, on the other hand, refer to grain direction as being **grain-short** (or cross grain) and **grain-long** (or with the grain). Paper is referred to as being grain-long if the grain runs parallel to the press cylinders (**Figure 4-4**). This usually means the grain runs in the long dimensional direction of the paper. For example, grain that runs parallel to the 38-in. (965-mm) dimension of a 25×38-in. (635×965-mm) sheet would be grain-long. It would be grain-short paper if the grain ran in the 25-in. (635-mm), or short, direction.

One way to determine the grain direction of paper is to gently place a piece of paper on the surface of water (it is important to keep the top

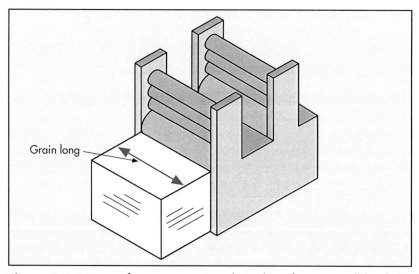

Grain long

Figure 4-4. Printers refer to paper grain as being long if it runs parallel with the face of the printing cylinders.

side of the paper dry). See **Figure 4-5.** As the paper floats on the water, the side in contact with the water will begin to wet and the paper will start to curl. The grain direction of the paper runs parallel to the curl.

Paper Finishes

Characteristics of the surface of paper are the result of the finishing process. You can observe the surface of a coated sheet, for example, by holding a sheet up to the light at an angle. The smoother the sheet, the smoother the reflection of the light on the surface. The smoothness of the surface will impact the ink lay.

Finish is the term used for the surface characteristics of a sheet. The paper mills use various methods to finish sheets to create a variety of looks or types of paper.

One finishing process is ***calendering,*** which is achieved by running the paper through a series of rollers called calender stacks. Calender stacks use a specific roller finish and count that, when combined with the appropriate amount of heat and pressure, yield the desired finished product. The extent that calendering makes an uncoated paper smoother depends on the number of rollers the paper passes through. The greater the number of rollers, the smoother the sheet will be. Calendering a coated stock works much the same way. Shorter stacks produce loose clay or matte sheets. Longer stacks produce dull stocks, and the longest stacks produce gloss stocks. Calendering has both positive and negative effects on paper. On the positive side, calendering makes the surface smoother and glossier, and it improves ink holdout. The down side is that the process requires heat and pressure, which, while improving the finish, can reduce the opacity and bulk.

Two-sidedness is a characteristic of paper that refers to the surface on one side of a sheet being different from the other. All sheets exhibit two-sidedness to some extent due to the configuration of the paper machines. On some blank sheets, the differences are subtle, while with others the differences are obvious. Paper machines have what are called wires and felts that are part of the process of manufacturing paper, and they impact the sheet differently. In many mills the variation in print quality between one side of a sheet and the other has been greatly reduced in recent years, lessening the impact of two-sidedness in coated and offset papers. To overcome the problem of two-sidedness, paper makers have evolved various configurations of ***twin-wire forming*** or "two-sided drainage" paper machines.

With some text stocks, sometimes the texture varies so greatly from one side of the sheet to the other that it needs to be considered in some designs. In a brochure with multiple pages, a pocket, or a flap, or any case where both sides of the stock will be visible at the same time, the difference in texture or color from one side to the other can be distracting. When consistency of surface is a concern, blank comps and/or printed samples of the stock should be used to help determine the potential impact, if any, before the stock is ordered.

Coated Paper Finishes

Coated papers have a surface coating that enhances the surface of the sheet and creates a smooth, ink-receptive surface. Coated papers come in a variety of finishes and qualities with distinct characteristics and uses.

Dull enamels are heavily coated and calendered with matte or cotton rollers to buff the surface and seal it without bringing up a gloss. One characteristic of a dull enamel is its low level of light reflection. Ink will "gloss" on the surface, offering a nice contrast to the finish. Small type is easier to read on the flat surface, and you can expect good ink holdout. Some sheets finished in this method are called by names other than dull, but the names still generally refer to the same flat, highly coated material. Dull finishes are less common in lower-quality grades because they mark more easily and there is less demand for them.

A *gloss enamel* paper in any grade is calendered with highly polished rollers under heat and pressure to create a reflective surface or gloss by polishing the clay in the coating. The ink gloss is higher on a gloss surface, and the holdout is very good. However, it may be difficult to read large blocks of copy because of the high gloss.

Matte finish refers to a coated material that is buffed but not sealed. The surface is a loose clay, so the ink holdout is not as good as with fully coated enamel. Matte sheets have a tactile softness, and the flat surface makes reading small type easier.

Cast coatings are made of special high-quality clays that render a finished surface of maximum reflectivity or gloss when the clays are applied to high-quality base stock and polished at a slow speed on a mirror-finish magnesium drum. These papers are available in a variety of configurations and weights. Some are coated on one side (C1S), and others are coated on two sides (C2S). These are always high-quality sheets, and the primary considerations when choosing from the available brands are the weight or thickness, whether the paper is coated on one side or two, color, brightness, printability, how well it folds, and the finishing processes necessary to complete your project.

Super premium sheets are produced using conventional coatings on premium base materials. The finished sheets can be either gloss or dull depending on the calendering. These sheets require high-quality coatings and longer finishing to achieve the surface characteristics for this classification. Due to the refined materials used in these sheets, relatively few brands are available in this category, and they are generally available only in a limited range of book and cover weights.

Of the **conventional coated grades,** the gloss and dull finishes are the most common surfaces for coated paper, and they are available in a wide range of qualities and weights. The least expensive of this type of sheet is the **#5 coated groundwood** paper. These stocks have higher opacity, lower brightness, and poor surface and formation quality. They are primarily used for publications, inserts, catalogs of lesser quality, and mass mailings. This material is usually run "web," which refers to a high-speed roll-fed press that usually prints on both sides of the stock at the same time and is equipped with a hot-air dryer.

The **#4 coated groundwood** paper has a higher percentage of chemical pulp mixed with the groundwood fiber. While this reduces the opacity, the sheet has a better surface, improved formation, and greater brightness. While the opacity is often lower in these sheets than in a #5, it is generally higher than in a free sheet. These grades are used for better publications, direct mail, and catalogs. The factors that generally determine the use of a #4 over a #5 are the quality of the color, the number of pieces to be produced, and budget. These are also primarily web sheets, but some mills do offer sheeted material in this grade.

The **#4 free sheet enamel** stock has either no groundwood pulp or only trace groundwood fibers in the pulp. They are a step up from the #4 groundwoods. These sheets are generally brighter than groundwood-content sheets, and while they do not have the same opacity as the groundwoods, they are more opaque than the other grades of free sheet enamels. These sheets are generally not as smooth and their formation is not as good. When running this grade of paper, it may be necessary to use a slightly heavier sheet to compensate for the lower opacity. This can make the finished piece feel more substantial, but it increases the paper and shipping costs. These sheets are generally available in both sheetfed and web grades.

The **#1, #2, and #3 enamels** are the most common grades of enamel free sheets and are used extensively for collateral printing. Every mill establishes its own quality standards for each classification within certain industry parameters. These are all fully coated enamels. The lower the grade numbers, the better the brightness, formation, surface, and color. These grades are available in a range of weights, usually book weights from 60 to 100 lb. and covers from 60 to 120 lb. Not all grades are available in all these weights.

The grade of a sheet is determined by the manufacturer using a range of factors that include brightness and opacity. Sheets in a given grade usually have similar ratings, but the grade is often just a guide to the cost and quality. Therefore, use samples of the actual stock being considered to make the final selection.

Common Uncoated Paper Finishes

Uncoated sheets come in as many varieties as coated paper, and the costs vary widely. There are premium uncoated sheets that have clay in the base stock, which enhances the ink holdout and creates a smooth, flat surface for consistent ink coverage. Uncoated sheets are available in a wide range of grades from commodity to archival-quality, pH-neutral sheets. In addition, the whites range from bright whites to neutral whites and soft whites. In the offset grades there are myriad colors and surfaces available, including a wide range of book weights and, in some lines, matching covers.

Vellum finish uncoated paper has received a minimum amount of calendering. Thus it is not a very smooth surface and will absorb more ink. However, it retains its bulk and has the highest opacity. Determining the grade of a sheet is done by each mill. Sheets generally fall into the same range of brightness and opacity factors as other sheets with similar grading, but the color, formation, and gloss can vary. Therefore, only use the grading as a guide for cost and quality.

Antique-finish uncoated paper that has been more highly calendered is smoother than the vellum and usually has better ink holdout, but its opacity is lower.

Wove finish is a smoother finish than the antique. These sheets have less bulk and opacity, but they have improved ink holdout.

Smooth finish papers are calendered to a high degree. They have better ink holdout than the other finishes. Because the opacity and bulk are lower, a heavier weight of this stock may need to be considered depending on the coverage and opacity requirements.

A wide variety of uncoated papers have embossed textures or patterns. Some are fairly common, like linen finishes. Others are unique to a particular brand. These papers are generally offset grades and can be found in mill swatch books or merchant sample books. They add variety to a printed piece but are generally not expensive.

Text stocks are sheets made with distinctive textures and finishes in heavier weights of book and cover. The finishes and colors are distinctive to each line. The colors are often richer and deeper in text stocks than in offsets, and they have matching cover weights usually of 65-lb., 80-lb., or heavier. The texture is generally achieved in one of two ways: wet texturing or dry texturing.

Wet texturing is a process whereby the fibers are formed into the pattern or texture on the wet end of the paper machine. Either felts or dandy rolls form these textures. *Text felts* are specially woven pieces of heavy material that are mounted on the paper machine; the pulp is pressed into the felt to take on the characteristics of the fabric. *Dandy rolls* are cylinders with copper wire woven into patterns. The roller is positioned on the wet end of the paper machine and rolls over the wet pulp, rearranging the fibers into the prescribed pattern. One of the most common uses of dandy rolls is creating watermarks in bond stocks. Some mills use both felts and dandy rolls to create specific textures. Wet-texture stocks are stronger because the fibers are not crushed into the texture. Generally the textures are more interesting and can be somewhat more irregular as the felts are woven, not formed.

Dry texturing is an offline process. This means that the texture is created in the paper after it comes off the paper machine. Sometimes these sheets are sandwiched between plates that carry the desired finish and pressure is applied. In other cases the texture is on a set of rollers that transfer the impression to the sheet as the paper passes between them.

Embossed textures are created by putting the finished paper through an embossing process. In this process, the pattern is built into a die and then pressed into one side of the sheet. This process creates a one-directional texture and can crush the fibers, which weakens the sheet. Dry and embossed textures tend to be more regular because they are less likely to have the irregularities in the texture created by impressing a texture in wet fibers. The offline finishing of text papers is less expensive and more common than the wet textures.

Bond and Writing Papers

The terms bond and writing papers are no longer used for the specific papers they originally described. Today, they refer to a general classification of stocks ordinarily used for letterheads, memos, and client-generated presentations. Bonds are not generally as opaque as offset or book papers. Writing papers are made with shorter fibers than regular bond stocks, making them softer and more receptive to writing inks.

Since most bonds and writing papers are used for letterhead or similar materials that need to be run through an imaging device, like a desktop printer, it is always a good idea to test the material in the specific printer if possible prior to producing the job. Some papers are not as receptive as others to the toners. Heat, for example, can affect some sheets. In addition, the feeder may limit the weight and finish of the stocks that can be used.

Following are some characteristics unique to these types of papers.

- *Rag content* refers to the amount of pure cotton fiber used in a sheet. You will find the rag content in the watermark of bonds and writing papers. Rag content is noted when at least 25% of the fiber used to make the sheet is from a cotton source. The higher the rag or cotton fiber content, the finer the stock.

- A *watermark* is an impression in the fibers of the paper made with a dandy roll on a paper machine. The dandy roll, a cylinder with a design and or type worked into it in a raised wire, rearranges the fibers while they are still on the wet end of the paper machine. Watermarks can be the name of the stock or custom-made for a specific company or client.

- *Sulfate bond* is a lightweight paper made from wood pulp with no rag content. These bonds are made in a range of qualities from a #1 to a #4. The lower the number, the better that the surface and print quality of the paper will be. Some mills make a #1 bond that is textured and watermarked for use as a letterhead stock in place of a rag content sheet.

Boards and Bristols

There are specific finishes and weights for each type of board (i.e., heavyweight paper also known as bristol board or bristol paper).

Tag stocks are usually very dense and durable and are used for things like file folders. Bristols come in finishes like vellum or plate, and they are frequently used for business cards.

Index stock is used for index cards and direct mail. These are generally lighter-weight board stocks and are referred to in pounds or plys.

Coated board stocks are usually referred to in caliper, points, or plys, rather than by weights. These bulky materials are rigid and durable and come in a wide variety of weights or thicknesses. These materials are used for point-of-sale (POS) materials like shelf talkers and neckers (cards that hang from the necks of bottles) because they print reasonably well, are strong enough to withstand the short-term abuse of a retail environment, and are generally economical.

Some boards are *solid bleached sulfite (SBS),* which means the material is formed as a solid sheet of white fiber. These boards are white on the edges and have a cleaner, finished look in the completed materials. The surface of these boards is often smoother and brighter than other types of boards. SBS boards can be coated one side or two, and while they are not as glossy as coated enamel, there is a slight sheen to the coated surface. SBS boards generally are available in weights from 6 pt. to 24 pt.

Multi-ply boards are made of layers of material that have been pasted together to achieve the final weight. These often have brown kraft pulp in the middle, with white material laminated on one or both sides. The edges are brown or gray, and they are generally heavier than SBS boards. Some of the multi-ply board is made for specialty uses, such as the weatherproof or water-resistant boards designed to be used outdoors.

Since there are so many ways that board weights are referenced (e.g., plies, caliper, and weight in pounds), securing samples of the weight you are requesting or the vendor is recommending is the best way to ensure that the material being used is what is expected and needed.

Paper Weights and Measures

Bulk, the thickness of a sheet of paper, is measured in thousandths of an inch (0.001 in.) using devices called **micrometers** or **calipers.** Book weight stocks are usually referred to by their basis weight, not in thousands since they are so thin, but you can measure the thickness of any sheet in this manner. If you know the caliper of a sheet, you can use it to determine the approximate thickness of a specific sheet count or to determine the sheet count of a specific thickness of that paper. For instance, you can divide the overall height of a stack of paper by the thickness of a single sheet of paper to determine the approximate number of sheets in that stack. Knowing a paper's bulk is especially critical in the production of books, where the thickness of the book's interior must be known before creating the spine for the cover.

One way to determine the approximate number of sheets in a stack is to do a measured count. First, count the number of sheets in a 1-in. (25.4-mm) stack, and then simply multiply the number of sheets per inch by the number of inches in the stack to determine the approximate number of sheets in the stack.

Cover weights are often referred to by both weight and thickness, while boards are usually referred to by thickness and number of plys.

Printing papers are manufactured in large parent rolls and cut into sheets for sheetfed printing or smaller rolls for web presses. **Standard size** or **stock size** refers to full-size sheets of a particular grade line and represents the sizes most commonly used. **Basic size** is the one size among the standard sizes used to establish basis weight (**Table 4-1**). Standard sizes that are stocked may vary from mill to mill, and between merchants. Offsets, book, and cover papers are usually available in a variety of standard sizes from the mill, while text covers tend to be more limited in standard sizes. Bonds and writing papers are routinely stocked in smaller sizes that accommodate letterhead sizes with minimum waste. Mill standard sizes are calculated on the traditional uses of the paper, so if you are using a paper in a nontraditional application you may have excess paper waste or need to consider a custom-size sheet.

Grade	Basic Size
Coated offset, book, label	25×38 in. (635×965 mm)
Coated cover	20×26 in. (508×660 mm)
Bond, business, duplicator, copy	17×22 in. (432×559 mm)
Vellum bristol	22½×28½ in. (572×724 mm)
Index	25½×30½ in. (648×775 mm)
Tag	24×36 in. (610×914 mm)

Table 4-1. The basic sheet sizes for several paper grade lines. *(Adapted from the Paper Buying Primer)*

Basis weight, however, is the weight, in pounds, of one ream of paper cut to its basic size. Internationally, the weight of all paper and paperboard is expressed as *grammage,* which is the weight in grams of one single sheet of paper whose area is one square meter (39.37×39.37 in.). *Ream weight* is the weight, in pounds, of one ream of paper. *M weight* is the weight of 1,000 sheets of paper.

A *Ream* is a packaging standard for paper. It is always 500 sheets regardless of the sheet size. Cut sheets are almost always ream-wrapped. Even some press sheet sizes are ream-wrapped in the cartons. Paper that is not ream-wrapped in a carton or on skids is usually marked in reams with slips of colored paper called *ream tags.* Ream tags are inserted in the end of the stack every 500 sheets to make counting easier. Heavier stocks like text covers are sometimes wrapped in smaller quantities for ease of handling but always in factors of 500. For instance a package of a text cover may be 125 sheets, so four packages equal a ream.

Pack-out is the amount of material in a carton or box or on a skid from the mill. This information is important for several reasons. Price breaks on small orders are determined by the packaging. For instance if you need three reams (1,500 sheets) of stock but the carton contains four reams (2,000 sheets), it might be more cost-effective to buy four reams at the carton price than three reams at the ream price.

5

The Mechanical

The Layout

When the creative concept has been approved and the client has signed off on the copy and images, it is time to begin preparing the file for production. The layout serves as a guide for the final mechanical. Sometimes elements of the layout can be reused and placed directly in the mechanical.

It is important to understand how a mechanical file differs from a layout file. For ease of alteration, layouts are loosely built. Although they may look fine when printed on a color laser, they are not generally production-ready. Ordinarily layouts do not have bleed, crop marks, or final images in them. In a layout, elements are sometimes placed over other elements, "cheater boxes" are used to hide parts of elements, and the art can be built using PANTONE MATCHING SYSTEM® (PMS) colors that will not be used in the production. Images in layout files are often boxed, not cropped, so you have more data in these image files than you need.

Building the Mechanical

Building a mechanical is an art form. When done correctly, it saves a tremendous amount of time and money. To build a mechanical, you must have all the elements of the project, as well as some specific information.

You first need to know the finished size of the piece, the bleed requirements, creep, and the safe area. With these dimensions, the form or *master* can be built with bleed, guides for safety margins, and crop marks. Masters (**Figure 5-1**) are very useful if you are doing multiple sheets or spreads where each page or spread is the same size and have common elements that should be positioned in the same place from sheet to sheet. For instance, in the case of 8½×11-in. sell sheets, the master sheet might contain repeating design elements, logos, or legal copy that appear in the same position on every version. Once an item is placed in the master, it appears on each page in the

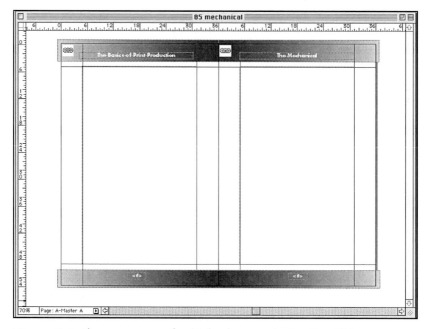

Figure 5-1. The master pages for this book, created using QuarkXPress.

same position. If there is an overall change or correction to that item, the change can be made in the master, and each occurrence of that element will be automatically updated.

A file might have several master pages. For example, master page A might be a chapter title page, and master page B might be a standard text page with running heads, folios, and text boxes.

Next you need to know the printing method to be used: e.g., digital output, screen printing, embroidery, or conventional printing. Where applicable, you also need to know what the line screen will be, how many colors should be in the final file, the PMS numbers if any, and the number of scans and their sources. With this information, files can be built for maximum efficiency.

Type of Output

Knowing how the material is to be produced and at what line screen dictates the amount of resolution that the images or scanned art must have to achieve the best result. Knowing this information allows the mechanical artist to avoid building files with too much or too little resolution. It can also determine in what format the file should be built. Some processes do not require as sophisticated a file setup as others. When a vendor receives a file that is built with his output

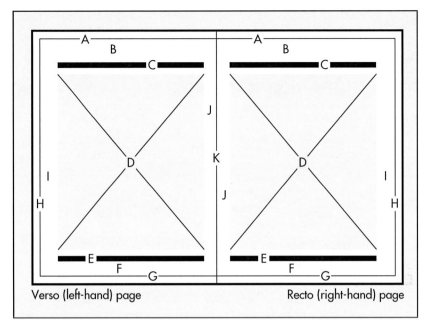

Verso (left-hand) page   Recto (right-hand) page

Figure 5-2. Basic page terminology: **(A)** head trim, **(B)** top, or head, margin, **(C)** header, **(D)** body of the page, **(E)** footer, **(F)** bottom, or foot, margin, **(G)** foot trim, **(H)** face, or front, trim, **(I)** margin at the face, or front margin, **(J)** margin at the gutter, or back margin, and **(K)** fold.

devices in mind, the files run better and there will be fewer charge-backs for file repairs or time delays due to disks being returned to be reconfigured.

Screen Mixes and PMS

In the mechanical file, all elements are set up to print in the correct screen mixes or PMS colors. When a layout is being built, there are often multiple versions of a screen mix, but in the final mechanical, all the screen mix reds, for instance, should have the same percentages of magenta and yellow. It is therefore necessary to check each area that should be the same color in the final and eliminate unused colors.

Some elements like logos and art are built using PMS or spot colors in the layout. If the piece is to be produced in four-color process, these colors will have to be converted to screen mixes. This is a simple process of telling the machine that the colors should be set to separate. This easy-to-do conversion is costly to fix if it is not caught until after the job has been prepped—or worse—is on the press.

Bleeds

Any image, color bar, background color, or other element that extends beyond the edge of the finished piece is called a ***bleed.*** This means that the printer must print that image or element past the trim marks to be sure that you have color to the edge of the finished piece. A printed piece can have a one-, two-, three-, or four-side bleed. (See **Figure 5-3.**)

Bleed must be taken into consideration when figuring the print area and paper size. While the amount of bleed required can vary, it is usually about an ⅛ in. (3 mm) on each side, but it can be as small as ¹⁄₁₆ in. (1.5 mm) or as much as ½ in. (12 mm). Therefore, before building the final files, always consult with the vendor producing the piece.

Bleed images should have the bleed built prior to placement in the mechanical so that they can be checked both for placement and trim prior to release to the vendor. Depending on the complexity of the image and the detail in the area that will bleed, it may be necessary to clone or duplicate the image area at the edge of the trim to create the bleed. Sometimes this needs to be done in an outside vendor's shop if the files are very large or the image area is complex or critical.

Flat color or screen mix bleeds should be built into the final files prior to releasing to the vendor unless the vendor has specifically stated that they will be added in his or her prep department.

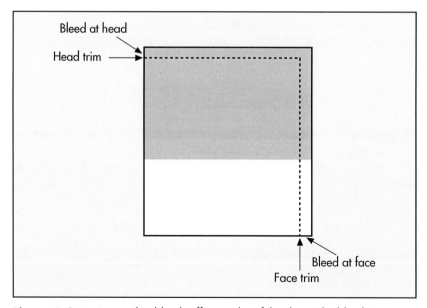

Figure 5-3. An image that bleeds off two sides of the sheet. The bleed is typically ⅛ in. (3 mm).

Trim Marks

Trim marks are short lines at right angles to one another outside the print area at all four corners of a flat printed piece. These marks indicate the trim or cut lines for the bindery. Except for bleed, all other elements should be placed within the trim marks with a margin for safety.

Folding Lines

Folding lines are single lines that indicate where a piece will fold and are placed outside the print area on each side of the piece in the direction of the fold. Like trim marks, in some cases you need to leave a little margin of safety on either side of a fold so type is not on the fold or an image is not crimped by the fold if it shifts slightly in finishing.

Live Area

Live area is the safe area inside the trim where all the type, art, and logos should fall to be sure that, when the piece is finished, nothing will be cut off or distorted. Determining safe placement is a matter of experience and mechanical limitation and should be done either by an experienced production person or by the vendor that will be producing the final piece. Measuring blank comps made to size and weight from the paper to be used in production can determine the creep—the slight, cumulative extension of the edges of each inserted spread or signature beyond the edge of the spread or signatue enclosing it—and trim. Live area guides are part of the guides the mechanical artist sets up in the initial page or master.

Die Lines

When a piece is going to be diecut either in a perimeter-cut shape or with a pocket or flap, a die line is needed to build the die. If there is a perforation, like a return card in the piece, a diecut hole in the front cover, business card slits in the back cover, or any other special processing of this type in finishing, a die line or keyline should be built in the file to indicate the exact position, size, and shape of the finishing required. On a die line or keyline, cutting lines are indicated with thin solid lines, perforations with dotted lines, and folds with thicker solid lines. When building these elements in the final file, it is recommended that the die line be designated as a "special color" which will not be output on the printing plate, but it should appear on the proof or laser to confirm cutting position and should be marked as "die line, do not print" as an added precaution.

Images

Scans or electronic files of the photographs or illustrations of the final images are color-corrected, retouched, cropped to size, scaled, rotated, and outlined if necessary and then imported into the final mechanical. They can be either the final high-resolution (high-res) files or resaved as 72-dpi low-resolution (low-res) files that will be automatically replaced by the high-resolution files at the vendor when the file is updated.

The image files should be scaled and cropped with image editing software (i.e., Photoshop) prior to being placed in the mechanical. This will not only reduce the size of the file but will insure that the image appears as it should.

In either Quark XPress, InDesign, or other page layout software, a picture box is drawn in the exact size and position you want the image to appear and the approved image file is placed in that box. Type can be embedded in the image in with the image editing software or placed over the image in the mechanical document. Once the file is finalized and approved, it can be flattened to avoid shifting, but this means that no changes can be made to the files. A copy of the layered files should be saved as well, so any changes or correction that might be necessary can be made.

Placed image files can be either the final high-resolution (high-res) files or if the images are not yet approved as final, 72-dpi low-resolution (low-res) files can be used as place holders and updated to the high-res images when the files are approved.

Final Copy

The approved and edited text to be used in the piece is also imported into the final mechanical, and appropriate type styles are applied to the text. Charts, diagrams, approved logos, and marks are also imported.

Necessary Copy

These are the elements that are required by the client or legal department or counsel to appear on the materials. They include the legal lines, company codes, copyright and registration information, trademark information, and dates. On layouts this information is often either not shown or its position is denoted by X's or greeking—the practice of inserting jibberish text as a placeholder to evaluate text and typography.

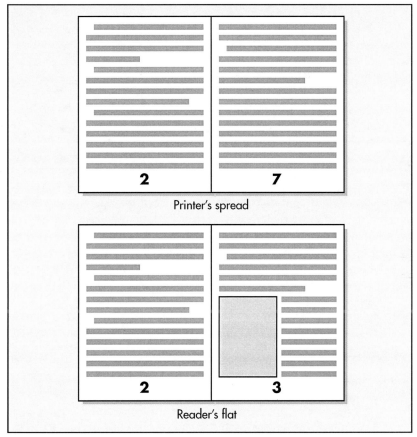

Figure 5-4. A printer's spread (top) and a reader's flat (bottom) in an eight-page booklet.

Reader's Flats and Printer's Spreads

Ultimately all folded materials must be laid out in ***printer's spreads*** that conform to the finished order of the book. A printer's spread (see **Figure 5-4**) is a pair of pages in the order necessary for printing, folding, and binding to yield the desired results. For an eight-page signature, pages 1 and 8 would be a printer's spread, as would pages 2 and 7, and so on. This can be done by the vendor from single-sheet mechanicals or ***reader's flats,*** which are just what the name implies—pages laid out in page sequence just as you would read them; e.g., pages 2 and 3 would be a reader's flat. Regardless of how the files are to be presented to the vendor it is important to have the live area and safety margins that we have already discussed built into the files.

If you are building printer's flats, request a blank comp in the specified stock for any multi-sheet book regardless of the binding configuration to confirm the page sequence. In stitched or perfect-bound materials, considerations such as the actual live area of each page are determined by the binding method. The blank comp should be made in the exact type and weight of paper and bound by the same method as the finished piece. The comp should be marked up as to page sequence and content. Then review the layouts and live area considerations with the vendor before the mechanicals are built.

Checking a Mechanical Prior to Release

Once the mechanical is built, you should review a desktop printout of the mechanical. To be sure you check all the client changes and notes, review the mechanical printout with the approved layout and mechanical instructions. Here are some of the things you should be looking for when checking a mechanical prior to release:

- Check the finished size of the piece. If the printout is not at 100%, ask at what percentage it was printed and confirm the proportions.
- Be sure the mechanical contains the approved final copy and that there are no typographical or grammatical errors.
- Check that all the elements are there and in the correct size and placement.
- Be sure that trim and fold lines are outside the print area and in the correct place, and the final size is accurate.
- Confirm that there is sufficient bleed, and make note of any areas where the vendor will be required to add bleed on the final printout.
- Check the shape and placement of any die lines/keylines that the piece requires. If there are perimeter cuts, make sure there is bleed or that the images extended outside the edge of those lines.
- Make sure nothing has drifted into the area between the trim marks and the live area.
- Output proofs and assemble them in the order of the book. Confirm that your pages will back up properly and that the sequences of numbers and content are correct.
- Check the image areas against the trim marks.
- If there are crossovers or gutter jumps (images or elements that run across the fold so they print on separate pages and come together in the binding), make sure they are in the same position relative to the trim marks in the layout and that the visual is correct when they are put together.

- Look at the copy and elements on either side of the folds. If anything is too close, now is the time to bring it up.

Reviewing a mechanical printout is very important. Having it read and reviewed by more than one person is advisable. However, remember that this is a mechanical of an approved design and the idea is to make sure it is right—not to redesign the piece.

The vendor will likely recheck the files in spreads prior to starting the final output, but you can save both time and money by taking the time to review the materials yourself.

Final Check of Files Prior to Downloading

Once all the type and position corrections have been made to the file, the mechanical artist should check the files for errors and oversights in the construction. Following are some of the items to check:

- Scan files for hidden boxes or layers that might cause problems.
- Inspect all elements to ensure that they have been rotated properly, scaled in the correct format, and cropped.
- Check all images and art to confirm that they have enough resolution at the output size for the line screen and that they have been converted to CMYK.
- Determine whether all unused colors in the palette have been removed.
- Inspect all spot colors used that do not print as PMS to ensure that they are set to separate.
- Confirm that all the screen mixes that should be the same values have the same name.
- Verify that all PMS colors being used are correctly identified.
- Confirm that the file is built for maximum efficiency in output.
- Determine whether redundant files have been eliminated.

If there are concerns about the setup, it is advisable to output separations on a laser to check the elements and values on each plate prior to downloading. This is an inexpensive way to avoid having errors in the film output. Once the mechanical file is ready, print out a full set of lasers from the final file(s).

Marking Up the Printout with Vendor Instructions

Once you have a clean, approved mechanical printout, it is time to mark up the vendor instructions.

The vendor uses this marked-up printout as a visual reference to the information on the digital file. The marked-up printout helps to confirm whether all the files are in place and whether the images are there. It also helps to verify that they have the correct number of colors in the final output. The marked-up printout of the mechanical should contain the following information:

- Finished dimensions.
- Total number of colors in the file. For example, a file that is set up for four-color process, PMS 185, and spot varnish would have a six-color output. On the other hand, the same file set up for four-color process, PMS 185, and an overall aqueous coating would only have a five-color output.
- PMS or match color elements identified through the use of a call-out. A line drawn from them to the edge of the printout where a color swatch or PMS chip should be attached.
- Any low-resolution images on the printout clearly marked "FPO, update with a high-res image prior to output" and the source of the high-res images indicated. For instance, say "images are on vendor's system" or "high-res images are on accompanying CD-ROM."
- Any bleed that needs to be added by the vendor. The areas requiring bleed should be marked and noted as "add bleed prior to output."
- Die lines/keylines clearly identified on the printout and marked "die line for position only, does not print" or "keyline for position only, does not print."
- Any special instructions for the printer, such as critical color areas, areas of concern like reverse type in an image, and so on. These concerns can then be evaluated prior to making plates or putting the job on press and corrected if necessary.
- A list of all the software and version numbers used to build the file.
- A list of type fonts used to build the file.

Once everything has been checked, the mechanical can be downloaded with all the support files and fonts to a disk. Ask the vendor which file transfer format is preferred, since there are so many options and not all vendors have the means of viewing each different type. A copy of the printed window listing all the files on the disk (**Figure 5-5**) should be inserted in the case or attached to the marked-up printouts. It is always advisable to open the files from the disk to verify that all the files are there and readable and that they are not corrupt.

When working with vendors chosen by the client or sending files to a client instead of a vendor, you may need to take at least one more step before you release the mechanicals. If the mechanical contains flattened files, it may be necessary to also send the layered files for those elements to the vendor for client changes. Discuss this with

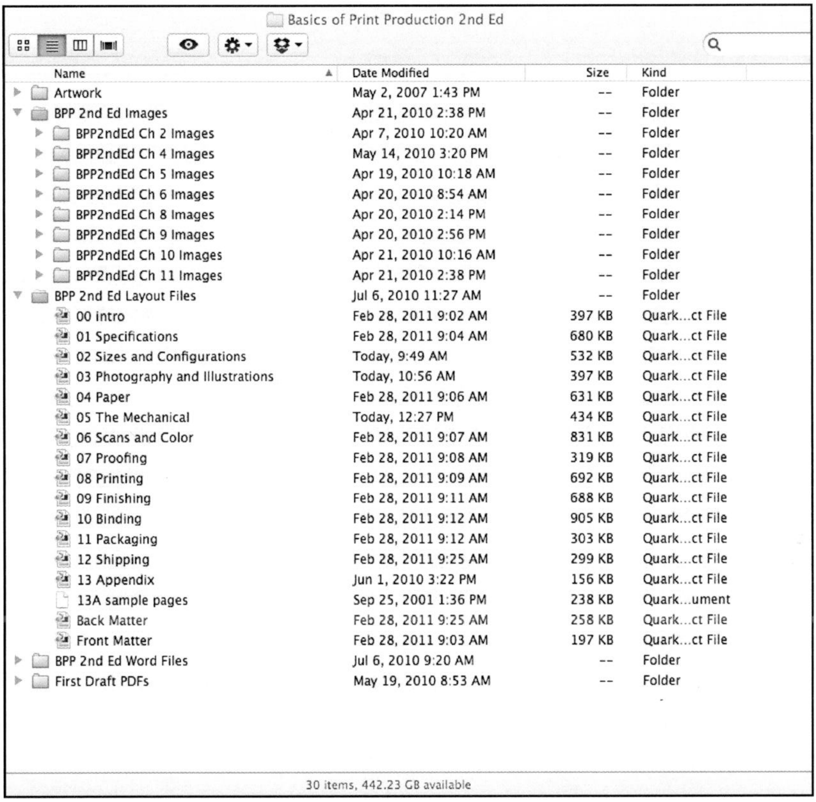

Figure 5-5. A screen capture of a window showing the files for this book

your client prior to shipping the files. If your client furnished you with only low-res images to build the mechanicals, be sure to note in your markups that you do not have the high-res files and that the client will need to update the files or furnish the vendor with the high-res images.

Send the marked-up printout, the printed window, the software list, and any loose color you may have for the images used in the files. It is advisable if your client has given you any color references that you give them to the vendor with the files so they can be taken into consideration from the start.

6

Scans and Color

Printing a black-and-white continuous-tone (photographic) image by most printing processes requires that the image is rendered into a grid-like pattern of dots. This process is called **halftoning.** At one time, most halftones were made on a large graphic arts, or process, camera. In this system, the photograph was mounted on a copyboard, and a halftone contact screen with a grid-like pattern of vignetted dots was placed over the film in the camera back. The light reflected off the photograph penetrated the vignetted dots to varying degrees, producing dots on the film corresponding to the tone values in the original photograph.

Today, most halftones are created digitally. Images are scanned (see the discussion of scanning later in this section), and the data is stored digitally. This data is often manipulated in an image-editing program, like Adobe Photoshop. When the image is output to final film on an imagesetter or printing plate on a computer-to-plate system, the raster image processor (RIP) creates the halftone dots as a series of spots (**Figure 6-1**).

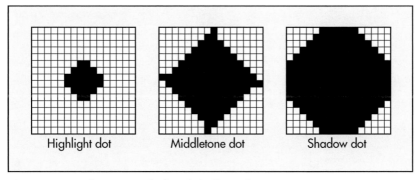

Figure 6-1. Enlargement of "digital" halftone dots.

Additive vs. Subtractive Color Reproduction

Despite all the ways that we think we have to reproduce color, there are only two basic methods—**additive** and **subtractive.** The additive process starts with black—the absence of light—and involves transmitted light before it is reflected by a substrate. The subtractive process starts with light already present and reflected from an object. Both processes are most commonly trichromatic; that is, they are based on the theory of using three primary colors to create all other colors.

Understanding the principles of the two systems is the foundation for understanding the many facets of the color reproduction process in printing. It's the basis for understanding tone reproduction, gray balance, and color correction—all crucial in achieving proper contrast, color balance, and color hue in halftone reproductions.

Additive color process (RGB). The additive color process involves transmitted light before it is reflected by a substrate. Adding and mixing the three primary wavelengths of light (red, green, and blue) in different combinations produces a full spectrum of colors. Adding all the primary colors in relatively equal amounts produces "white" light. Computer monitors, television screens, projection TV, and stage lighting are based on additive color.

Mixing any two of the additive primary colors will always produce another color, called a "secondary color." The secondary colors in the additive process are cyan, magenta, and yellow—which are the primary colors of the subtractive process.

Subtractive color process (CMY). The subtractive color process is based on light reflected from an object and which has passed through pigments or dyes that absorb or "subtract" certain wavelengths, allowing others to be reflected. The primary subtractive colors—cyan, magenta, and yellow—can be combined to form red, green, and blue as secondary colors. Combining the ideal primaries in equal amounts produces black.

The subtractive color process is what allows us to see color in the objects around us. A green ball, for example, appears green in white light because the colorants in the ball absorb the red and blue wavelengths and reflect the green. Of course, in a light source that is minus a green wavelength, the ball would appear black because there would be no green wavelength for the ball to reflect back.

A source of confusion about color can be traced to the way most of us learned about "primary" colors—with a box of crayons in grade school, where we were taught that the "primary colors" are red, blue, and yellow. Many people in the printing industry, especially in the pressroom and in interactions with clients, perpetuate this error by calling cyan "blue" and by calling magenta "red."

The actual printing of a full-color image uses the three process colors—cyan, magenta, and yellow—and black (CMYK) in various combinations of microscopic halftone dots to produce a myriad of colors, referred to as the **color gamut.** The client should realize that the gamut of color produced by any of the printing processes (thousands of colors) is significantly smaller than the gamut of a computer monitor (millions of colors), meaning that some colors will be difficult or impossible to reproduce.

If you look closely at the printed halftone dots with a magnifier, you will see that only eight colors are printed on the paper: cyan, magenta, yellow, black (black or overprint of cyan, magenta, and yellow), white (unprinted paper), red (overprint of magenta and yellow), green (overprint of yellow and cyan), and blue (overprint of magenta and cyan). At a normal viewing distance, however, the eye interprets the tones created by the varying combinations of the process colors as a full-color image (**Figure 6-2** and **Figure 6-3**).

Process screen mixes are used to not only recreate full-color images like photographs and illustrations in the printed piece but also to create any element or background that is to be printed. The entire element is broken down into the specific color combinations required to create each color element in the piece (**Figure 6-4**).

Figure 6-2. An enlargement of a "full-color" image, showing that there really are only eight colors being printed: cyan, magenta, yellow, black, red, green, blue, and white. (From *Color and Its Reproduction* by Gary G. Field, Printing Industries Press.)

Figure 6-3. An enlarged section of a photomechanical color reproduction. By gradually increasing the viewing distance, the human eye begins to perceive a single color instead of resolving the separate color elements that form the color. (From *Color and Its Reproduction* by Gary G. Field, Printing Industries Press.)

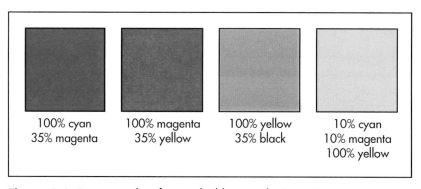

| 100% cyan
35% magenta | 100% magenta
35% yellow | 100% yellow
35% black | 10% cyan
10% magenta
100% yellow |

Figure 6-4. Four examples of screen builds created using various percentages (tints) of cyan, magenta, yellow, and black.

When the color data is formatted for the press, the dots of each color are aligned at a specific **screen angle** that is designed to lessen the appearance of moiré patterns in the printed images (**Figure 6-5** and **Figure 6-6**). If the colors are at the proper angles, a rosette will appear.

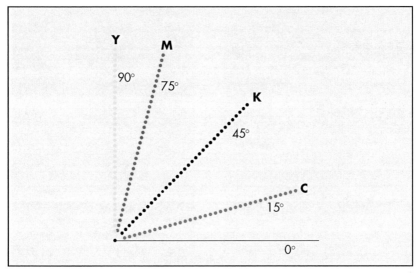

Figure 6-5. The most commonly used screen angles for four-color process work, keeping a 30° separation between colors likely to cause moiré. (From *Understanding Digital Color* by Phil Green, Printing Industries Press.)

Figure 6-6. Incorrect halftone screen angles will cause moiré patterns in certain colors: incorrect angles (left), and correct angles (right). (From Color and Its Reproduction by Gary G. Field, Printing Industries Press.)

Figure 6-7. The same image reproduced at different screen rulings: 150 lines per inch (left) and 65 lpi (right). (From *Color and Its Reproduction* by Gary G. Field, Printing Industries Press. Photo courtesy of ISO.)

The resolution or fineness of the ***line screen*** at which the piece is to print determines the size of the dots. Line screen, also called ***screen ruling,*** is the number of dots per linear inch to be printed. The screen ruling is determined with the help of the printer by taking into consideration the printing process, the press, the paper or other substrate on which the piece will be produced, and the design. (See **Figure 6-7.**)

A process called ***stochastic screening*** eliminates the use of screen angles. Instead, this halftoning method creates the illusion of tones by varying the number (frequency) of micro-sized spots in a small area. Unlike conventional digital halftoning (**Figure 6-1**), the spots are not positioned in a grid-like pattern. Instead, the placement of each spot is determined as a result of a complex algorithm that statistically evaluates and distributes spots under a fixed set of parameters; to the casual viewer, it appears that the spots are randomly positioned (**Figure 6-8**). This process avoids moiré and, in some instances, works well for detail and soft tones. Check with your vendor regarding the availability of stochastic screening as an output option. Your vendor may recommend this process for a specific project, such as a job being printed using HiFi color (e.g., six- or seven-color printing with two or three special inks in addition to the standard cyan, magenta, yellow, and black).

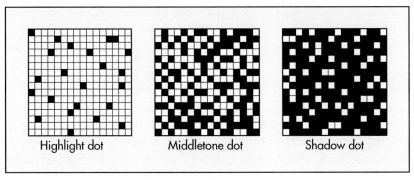

Highlight dot   Middletone dot   Shadow dot

Figure 6-8. Enlargement of stochastic dots.

Scanning

With the advent of digital cameras and computer-generated illustrations, the number of files that have to be scanned and converted into digital-format electronic files has been greatly reduced. But in the event that it is necessary to convert hard copy, the process is as follows.

The art is placed on a ***scanner,*** which is a device that uses light and filters to convert the image into an electronic file that contains color channels. Generally images are scanned as RGB or black and white, then the mode is changed for the required output.

Once the scan is complete there are various modes that can be selected for the image based on how it is to be used. If the image is going to print in one color, it will need only one channel to print as a black-and-white halftone. If the file is going to print full color, then the mode would be CMYK, and the image would have four process-color channels.

Sometimes an image will require more than four channels of color information. For example, if the image will be printed using high-fidelity color, as many as eight color channels may be required, one channel for each color of ink that will be used on the printing press. Some images include elements that are a ***critical color*** that can only be achieved by adding a Pantone or custom color to the separation. These images are manipulated so there is both four-color process and the match color within the separated file.

When scanning is required to prepare an image for printing, several pieces of information need to be given to the vendor when releasing elements to be scanned:

- **Type of scans needed.** Specify the type of scans to be made using the end product as the reference, such as four-color process, black-and-white (grayscale) halftone, duotone, tritone, and quadra-tone. Images that will ultimately be printed using the four process colors are usually scanned in the RGB mode and then manipulated as required in an image-editing software program. Images that will be printed as black-and-white halftones, duotones, tritones, and quadtones are usually scanned in grayscale mode.

- **Halftone screen ruling.** Specify the line screen at which the piece will be printed to be sure that there is sufficient resolution in the image. The resolution of the scan can be adjusted slightly up or down if the line screen or final size of the image changes after the scan is done, but if you have radical changes it is often best to rescan the image.

- **Image size.** Indicate the finished size of each element and whether or not it will bleed. This specification, along with the halftone screen ruling and required resolution, determines the size of the final file. Do not scrimp on the resolution to reduce costs, but don't overestimate the required resolution either, because too much resolution will make the size so large that it is difficult to work with the file.

- **Crop.** Crop is the portion of the image that you will be using in the printed piece. Mark the area for width and height in the correct proportion. If only one direction is critical, mark that area and tell the vendor the other dimension can fall where it may. If the image bleeds, be sure to indicate whether your crop is live or includes the bleed.

Types of Scanners

Inexpensive scanners that lack a wide range of color control features are often used to produce low-resolution "for-position-only" (FPO) images that will be used in the layout to show the client how the finished piece will look and as guides to show the vendor the cropping and placement of images. A high-resolution scan ordered from a vendor, scanned at the correct size and resolution, will need to be placed in the mechanical and used for final output. There are basically two types of electronic scanners used for this purpose: the flatbed scanner and the drum scanner. Both types of scanners are available in a wide range of sizes as well as the quality of the scans they can deliver.

The *flatbed scanner* is just what it sounds like. The reflective image is placed flat on the glass, the lid is closed, and a light source illuminates the image. In a single pass or a series of passes, the image is recorded as a digital file using either RGB or CMYK as the recording mode. The size of the original that can be scanned is limited by the size of the image area of the bed of the scanner and the bulk of the piece. Flatbed scanners are usually somewhat slower than drum scanners, but this format makes it possible to scan materials that cannot be scanned on a drum. Although some people consider the quality of scans from a flatbed scanner to be slightly inferior, advancements in flatbed scanning technology have improved the quality.

A *drum scanner* has a removable glass cylinder on which a transparent image is mounted using gels or oil to keep the image flat on the surface of the drum without air pockets. The cylinder is then mounted on the scanner. As the cylinder spins at high speed, a light source inside the cylinder illuminates the image. A slowly moving recording head on the outside of the cylinder captures the image data in a digital file or outputs a set of film separations directly. The width and circumference of the cylinder limit the size of images, but if you are working with transparencies or negatives, it is unlikely that you will encounter a problem locating a drum scanner able to accommodate your materials.

In addition to being small enough to fit around the drum, artwork or illustrations for drum scanning must be flexible enough to wrap around the drum. If the board is not flexible, it may be peelable, which means that the top layer of the board can be pulled away from the base, leaving you with a much thinner piece of material that can be wrapped on the scanner drum. There are some concerns when peeling a board, however. If the medium used in the creation of the art is too thick or brittle, it can crack or flake off when the board is peeled. In addition, some vendors will require that the board be peeled before they receive it, as they do not want to be responsible for any damage to the art. If the board is not wrapable, cannot be peeled, or is too large for a flatbed scanner, a photograph of the art may need to be made and the scan done from the transparency. Any time separations are made from something other than the original art or transparency, the color from the scan may not be as true to the original, but you can enhance the color using the pre-scan setting on the scanner or using color correction to the digital file.

After the Scan: Checking Loose Scans and System Work

Once the scan is done and the image is in digital format, the vendor makes loose color proofs. This give you an opportunity to look at each image and evaluate the contrast and sharpness and also to indicate whether color adjustments are required. While you are checking the loose color proofs, also inspect the images to see if any cleanup is necessary. Invariably, you will find little imperfections, specks or fibers in the images on the proof. These imperfections may only appear on the proof, but then again, they may be something that was picked up in the scan. All you need to do is circle these imperfections; if they are actually in the file, the vendor will clean them up.

In addition to the color, contrast, and cleanups, several things may need to be done to a scanned image. Reworking an image to remove or add elements is referred to as *system work.* The cost of these additional processes is determined by the amount of time, the degree of difficulty, and the equipment required to do them. *Retouching* is the process of removing flaws in the image, like smoothing out the edges of a product, toning down a light reflection that is creating a hot spot, or lightening a shadow on an element to bring it out from the background. Removing an element from an image and replacing it with the background texture of the image is called *cloning.* Making the background appear to be part of the original image can be fairly simple or can be very delicate and expensive depending on the image. *Pathing* is the process of outlining an element so that it can be used alone without a background or placed in a background or another image. Creating a path can be fairly straightforward if the element is sharp with defined edges like a bottle. However, if the element is soft and ill-defined like hair or fur, making the pathed image look natural requires working at very high magnification and can be time-consuming. The complexity of the path determines the time and cost involved. Taking elements of two images and creating a composite image with them is called *merging images.* Again the complexity of blending the elements into one so that it appears to be a single image, not one image pasted on another, determines the cost.

Elements that are in an electronic file can be converted to the appropriate combination of process-color screens either by choosing a PMS color from the color picker and telling the computer to separate it to process, or by making a custom color and entering the exact amount of each of the process colors you want in that color.

Beyond Four-Color...

You are not limited to four-color process. You can have a file built that has a **touch plate.** This is a PMS or custom color that is incorporated into the process to add emphasis or to make an element the exact color the client wants.

Systems are also available that have special software for creating six-, seven-, or eight-color process files. This is called high-fidelity color. Depending on the system being used, these additional colors are either fixed by the system or you can choose the additional colors to best accent your image.

When producing an image that is not four-color, you may want to increase the impact of the image by adding tone and detail, like a sepia effect for instance. This can be done using four-color process printing or you can make the image a duotone. A **duotone** is an image produced using two colors of ink. The inks can be process color or PMS color. Duotones have more detail and depth than black-and-white or monochrome halftones. In addition, if you do not have any four-color process in the piece, it is less expensive to use two PMS colors or a PMS and black. An image that uses three colors of ink to create subtle tones is a **tritone.**

7

Proofing

Proofing is generally considered the last chance to check the piece for errors prior to plating and printing. Color proofing is an indication of the way your color will appear on the printed piece. How accurate the proofs are to the printed result depends on the type of proof and how well the system is calibrated to the press.

Loose Proofs vs. Random Proofs

When working with a high-resolution image that has not been color corrected, a proof of the image needs to be made. The purpose is to see the color and details in the image to determine what, if anything, needs to be done to correct the image and color before it is placed in the file with all the other elements. This is called a ***loose color proof.*** Pulling loose proofs allows you to make smaller proofs, which are more cost-effective. The changes and color correction can be made to the image file before it is placed into the mechanical which saves time and helps to avoid having the wrong file in the final mechanical.

The choice of loose proofs, while not necessarily what will be used for the final composite proofs, can be influenced by the final proof configurations. Rather than have an individual proof of each image, many vendors will gang up several images on one proof; this type of proof is often referred to as a ***random proof*** or ***scatter proof.***

With the approval of the loose color, the mechanical is updated with the high-resolution image files. Once the approved file is complete with high-resolution images in position, a ***composite proof*** is made as a final check that all of the elements are in place and the color is accurate. Composite proofs are just what the name implies: a visual of the piece with all of the elements, copy, and art in place to final size and color, just as the piece will print. Reduced-size composite proofs are used for some oversized projects. In that case all of the elements are proportionate to one another in the output relative to the final size.

Proofing Methods

There are a variety of proofing options from press proofs to digital to virtual. Some factors to consider when choosing which type of proofs to use include color accuracy, sharpness, cost, and time.

Press Proofs

The most accurate proof would be a ***press proof,*** because it is actually printed on a press. To create a press proof, the files are plated and printed, sometimes with the client there. The press sheets are registered, color balanced, and the job is pulled off the press. The printed sheets—i.e., press proofs—are then sent out for approval just like any other proof. You can also run one color at a time to see the density or concentration of each color in any given area or run the colors in sequence to see the impact of one color on the next. When you pull single-color and sequence sheets, the proofs are called ***progressives*** or ***progs.*** You can do a press proof without doing progs, and not all press proof quotes will include them.

Generally, whenever a press proof is going to be made, a composite proof is made from the film prior to plating. This additional step is used to ensure that the color is close and that all the elements are there before going to the expense of doing the press proof.

Press proofing is usually performed when the color is critical and approval needs to be done by someone unavailable to go to the actual pressrun. Once the color on the press proof is up and in balance and approved by the person charged with checking the press proof, a sheet is sent out for final approval. After the sheet is approved or marked with corrections by the client, it is returned to the vendor. If the sheet is approved as is, the job is ready to be printed. However, if changes are requested, new film and plates may need to be made, and the magnitude of these changes could require a new round of press proofs.

You would not use press proofing for loose color for a number of reasons. First, since all the elements that will appear on the actual press sheet are not in place, the ink takeoff and color impact are not the same, meaning you would not get as accurate a proof. In addition, there are time and cost factors to consider.

Press proofs are not done for a large percentage of printed jobs, but there are clients who require them and jobs that need them. In some instances, press proofing can save both time and money by isolating problems prior to the actual run that cannot be identified by conventional or digital proofing means. Press proofs are the most accurate method of proofing, and in some cases, they are well worth the time and cost.

Conventional Proofs (Film Proofs)

At one time, proofs made from film were the most common method of viewing loose color and composite proofs. The process was refined to the point where the proofs reflected color, very close to what was achievable on press, so when plates were made from film, these proofs were the most accurate off-press proofing system available.

With the transition to direct-to-plate technology, film became obsolete and is no longer available in most printing and prep facilities.

Overlay proofs, such as *color keys,* are one of the oldest types of film proofs and now have a digital equivalent. Each color can be generated on a separate sheet and then aligned to one another in register to create the final visual.

As with film each sheet contains only the part of the image/copy/design that will print in that color, just like the image on the printing plate. So if the job were going to be two colors, then two sheets would be output, one for each color. For four-color, four sheets would be output.

The loose sheets, representing each color put together—with marks aligned to ensure the piece is "in register" and the colors fit together exactly—are then affixed to a sheet of stock, often the actual stock on which the job is to be printed.

When complete, the layered sheets show how the image will appear in an approximation of the finished color and can be used to confirm color breaks and fit as well. If a project is not being printed in four-color but using PMS or match color, this method will often give the best color match available.

Overlay proofs are also used to show how color printing will appear on color stocks, on non-paper substrates like foil or plastic, for two- and three-color jobs, and occasionally for process work. Overlay proofs are not used for loose color proofs, but they are still used for composite proofs. Overlay proofs and progressive press proofs are the only proofing methods that allow you to look at each color alone or in specific combination on the proofs. Most flexographic printers and screen printers still use overlay proofs frequently.

Integral film proofs are proofs in which all the colors are bonded to a single base sheet. Several methods of making integral film proofs are available. In one system, a clear film is laminated to the base sheet. The combination is exposed to the film for the yellow printer and peeled, leaving a sticky, clear image area on the base. The sheet is fed through the yellow toning cassette of the toner-processing unit, and the yellow powdered toner adheres to the sticky image. These steps are repeated for magenta, cyan, and black. This system can be customized to the printing ink color by blending toners to match the hue and by varying the speed at which the sheet goes through the processor to alter the toner density.

With an integral film proof, the result after all four colors have been processed is a single sheet of material that is a fairly accurate representation of the image for color, sharpness, and contrast. The film must be placed in exact register from one color to the next.

You can also proof Pantone and other spot colors in this manner. Some systems have a fairly wide range of stock colors, custom colors can be ordered, and some systems allow you to mix colors. This is very important if you have a touch plate or a fifth color being used in combination with the four process colors.

The base material used in the proof is selected to mimic the amount of dot gain that you can expect on the press. In addition, different base whites can be selected with some systems to better indicate the impact of the paper stock on color. Some film proofing systems even allow you to proof on the exact paper being used to print the project. While this is an excellent way to see the impact of the color and texture, there is no way to duplicate the amount of ink absorption by the paper.

Digital Proofs

Digital proofs are generated directly from the digital file to a proofing system that images the colors onto specially treated materials using electronic imaging methods. The sizes, accuracy, and reliability of these systems are as varied as the manufacturers that offer them. Some of these devices are so well calibrated that they are extremely accurate compared to the color you will get on the film, while others are strictly for proofing content and position and should not be used to determine final color. Digital proofs are usually less expensive and less time-consuming to make than film-based proofs.

The best way to determine if a digital proof is accurate enough for your purposes is to ask your vendor if they have done *test comparison proofs.* These are proofs of the same image file made digitally and from film.

While color accuracy in digital proofing has improved greatly in recent years, there are still some things to be aware of. For instance, depending on how a file is built, digital proofs can either not show a spot color or the color will be created using the four process colors so it is a screen mix and—depending on the color—may not be very accurate.

For instance, if you are going to print a two-color piece containing a duotone of black and purple, and you proof the job on a four-color digital proofing system, the purple will be created using four-color process screen mixes, and the proof may not necessarily be a good representation of how that image will look when printed, so it is a good idea to weigh all the parameters of your project before deciding on a proofing method or relying on the proof as an accurate representation of how the printed piece will look.

Advancements in digital proofing have been impressive, and it has become the preferred method of proofing for most shops and most projects. The proof sizes that originally were a maximum of 11×17-in. plus bleed have been expanded and continue to grow. There are a variety of sizes and capabilities available for a number of equipment manufacturers. Digital proofs can be calibrated to the SWOP standard, to a specific press gain curve, or to a custom color formula. In addition to this, digital proofs can be run on production stock, the paper the job will print on, or mimic the properties of the publication.

There are so many variables and options available in digital proofing, and they are changing and growing all the time. You should talk to your vendor about their proofing options, capabilities, and size and let them help you determine the best fit for your project.

Virtual Proofs

As the name implies, this is a totally electronic form of proofing. For the proof to be as accurate as possible:

- A specific system will need to be designated for viewing all proofs
- Calibration of the receiving monitor is required
- The viewing area needs to be in an environment where the lighting is both controllable and consistent around the system.

The goal of this system is to transform a monitor into a color-managed, electronic light booth. If the color accuracy proves out, the timesavings and logistics that this option could address would be impressive. But unlike a traditional light booth, you could not view just any file for color, only files that contain the color calibration information and are compatible with the virtual proofing software would work—and that software is usually proprietary to the vendor who installed it.

Once the system is set up to receive the "proofing" files, the vendor posts the files along with a color control packet that adjusts the visual to reflect the properties set up in the proof, custom formula, SWOP/GRACoL standard, which downloads with the image and sets the monitor to those settings when opening the proof file.

In order to be sure you know what you are looking at, it is a good idea to use a set up file for which you have a color-correct hard proof, a printed piece, or both available to reference along with the virtual proof on screen to see how it compares. It is sometimes hard to adjust for impact of the backlit screen on images with hot highlights and high-key colors.

Proofing Considerations with Computer-to-Plate Technology

The introduction of platemaking technology where plates are imaged directly from the digital file by lasers, called ***computer-to-plate (CTP),*** raised a number of questions with regard to proofing final files.

Since there is rarely the option of film proofs, and if digital proofing, even with the advancements, is not accurate enough or you are running custom color, you might choose to run a press proof.

With a critical color job or a job that has very specific press issues, a press proof is the best way to proof and the only way to address not just the color but how it will run. This can be expensive, but not as costly as putting the job on press and having it not work while you are paying for the press time. Or, worse yet, the job may be run and then rejected because of color issues resulting in the need to do a reprint.

8
Printing

Planning

Just as with any project, a plan needs to be developed for a printed piece to determine exactly what will be required to produce it, the associated costs, and possible pitfalls. Constructing a print plan is usually required to do a quote. The plan determines such things as the press sheet size, the number of sheets, the number of passes through the press, ink requirements, the finishing and bindery requirements, and packaging. Once this information is determined, it is possible to estimate the cost and the time requirements for a specific project.

Several factors are taken into consideration when planning and quoting a job, including the following:

1. Sizes and number of elements, or printed pieces, for efficient use of paper stock.
2. Grain direction of the stock for folding and rigidity.
3. Quantities of each element if they are not common.
4. The amount of ink coverage and elements of the design for best ink takeoff, to avoid ghosting, and to give optimum color control on press.
5. Finishing requirements—if only one element needs to be diecut, for instance, you need to be sure it is positioned to ensure the most accurate diecutting, or it may need to run on a separate form.

With these factors in mind, the printer can begin the estimating process for the following.

Paper Stock Requirements

When determining the amount of paper required to print a job, more than just the final quantity must be considered. Each process that the sheet undergoes during printing has two additional considerations: *makeready* (the paper required to set up a process) and *spoilage* (the percentage of waste anticipated in each process). Allowances for these two types of waste must be made when developing the plan for a job. Therefore, each step of the process adds to the amount of overrun the printer must factor in to be sure that count is achieved when the job is complete. Adding up all of these projected requirements establishes the approximate amount of paper that will be needed to print the job.

Another factor that can impact the total cost of the job is the sheet sizes. If the specified stock is not available in an efficient-size **parent sheet** for the job, it can mean more waste if there is either not enough time or an insufficient quantity to order a custom sheet size.

Grain Direction

As we discussed in the paper section, grain direction in some papers is very important to the way the finished piece folds as well as the feel of the piece. Grain direction is also important in the way a sheet prints and the amount of stretch or fanning you might encounter on press. Since only a finite number of parent-size sheets are available and custom orders are not always feasible, printers sometimes cut down parent sheets to fit the job. However, doing so may change the grain direction of the sheet for the press. So when a job is planned, if there is a critical color fit, folding considerations, and so on, it may be necessary to use a slightly larger standard-size sheet with the correct grain direction, and that has a cost impact.

Multiple Press Passes

The number of units on a press limits the number of colors that you can print in one pass. So obviously, if the design calls for eight colors on two sides and the press only has six printing units, the piece would have to be run through twice per side to get all the colors printed.

Dry Trap

Sometimes the plan will call for multiple passes, not because of the number of colors but for a technical reason or effect. When you print a second pass, the ink is **dry trapped,** which means you print wet ink over dry ink instead of printing everything in one pass, which is called **wet trap.**

Dry trapping is not always optional. Sometimes equipment limitations and ink properties dictate a dry trap, while other times design parameters, time, budget, and critical registration dictate wet trapping.

Press Form Layout

The printer's planning department will determine the layout using the factors just discussed as well as their equipment limitations.

Sometimes a job can be printed with a front and back on the same side of the press sheet so only one set of plates is required; this is called **work-and-turn** or **work-and-tumble,** depending on whether the paper gripper edge changes when the second side of the press sheet is printed (**Figure 8-1** and **Figure 8-2**). You can perform a press check and do color approval on both faces of the piece at one time. When the sheet is turned over, it prints the exact same images

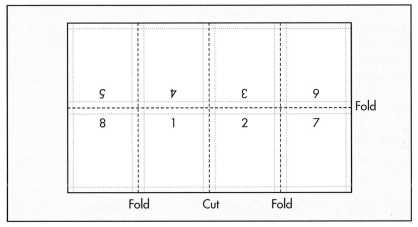

Figure 8-1. An eight-page work-and-turn layout. The first side is printed, and then the paper is turned over from left to right so that the same gripper edge of the paper will be used when the second side is printed.

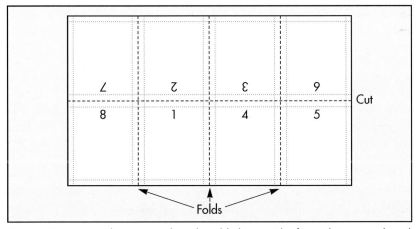

Figure 8-2. An eight-page work-and-tumble layout. The first side is printed, and then the paper is turned over from left to right so that the same side edge of the paper will be used when the second side is printed. Since a different gripper edge is used for the second side, the squareness of the paper is extremely important.

with the same plates, but they are backing up the opposite position from the first pass. This can save both time and money, but it is not the best choice for all jobs.

When a press sheet is laid out to print fronts on one side of the press sheet and backs on the other, this is referred to as **sheetwise** or **work-and-back.** Two sets of plates will be required, and you will need to do a press check and color approval for each side.

Perfecting is the process of printing both sides of the press sheet in one pass. Most web offset presses and a growing number of sheetfed offset presses have this capability. Perfecting is often cost-efficient and saves time because the front and back of the form can be press-checked and color-okayed at this time. In addition, the **backup,** which is the alignment of the two sides to one another, can also be checked at this time. The vendor will determine if a job is suited to this process and lay it out accordingly.

How Printing Works

On a conventional press, printing inks are used and they work much like the gels used to create colored spotlights in the theater. A wide array of colors is created from just a few gels by adjusting the intensity of each spot and overlaying their beams. The effect is much the same in printing.

In **process-color,** or **four-color printing,** various combinations and dot percentages of cyan ink, magenta ink, and yellow ink—the three primary colors in subtractive color theory—and black ink are used to emulate most of the colors in the visible spectrum. The percentages—or screen mixes—of the colors contained in an image are determined in the separations. The exact percentage of a color to make up a tone is converted to halftone dots set at a specific angle and then recorded on the film and plates. With the exception of black ink, which is opaque, the process-color printing inks that are applied to a press sheet act basically as microscopic transparent colored filters that absorb certain wavelengths of light and allow other wavelengths to pass through to the paper and be reflected back to the viewer. It is the combinations of the non-absorbed wavelengths that creates the range of color seen by our eyes.

Since light is such an important factor in printing, it is impacted by various factors in the paper: the smoother the surface of the stock and the better the ink holdout, the purer, brighter, and crisper the color will be. As we will discuss in the various printing methods, when toners are used instead of ink, the colors will not be as crisp because the color laid down by toner is not as smooth and transparent as ink.

Some colors cannot be duplicated in four-color printing, but generally you can achieve an approximation of those colors. If there are critical colors in your project that must be exact matches like a corporate ID color, PMS, or match color, inks can be used either alone or in combination with the process-color inks. At this point digital presses do not offer the option of running a job in all PMS colors, so if your project is designed with multiple PMS colors with or without process, it will have to run on either a conventional lithographic press, a screen printing press, a gravure press, or a flexographic press.

Many factors can influence the visual appearance of the colors in the print. For example, the color of the paper affects the color of the print because of its impact on the light reflection. Also affecting the visual appearance of the color is the surface smoothness or texture of the paper and how well the ink or toners set up on the surface of the paper. Another factor that can influence the way an image prints is its position on the sheet and the color elements located directly above and/or below it on the press sheet. All of these factors are taken into consideration when the job is laid out for press.

Types of Printing

Printing is no longer just a matter of choosing the right press. It now requires determining the best method of getting your project on paper. Because of the advances in the digital printing area, we now have options that make it possible to produce good-quality, short-run four-color materials—sometimes at a fraction of the cost of conventional printing and often in a much shorter time frame.

Digital Presses

Digital printing has become a dynamic and versatile option for the production of four-color printed materials. While there are still limitations to the process, the benefits often outweigh them for many projects.

Generally, digital presses (*Figure 8-3*) are ideal for short run, four-color projects that fit the size and paper stock limitations of the press you are using. While digital presses do have some size limitations, it should also be noted that digital web and sheetfed presses are being made available in larger sizes. They can run a wide selection of stocks and substrates, and the print quality continues to improve.

There are also some unusual options available. In digital, including large-format—where the size is determined by the width of the material and can run to virtually any length—the number of feet on the roll is the only limitation if you do not want a seam.

Be aware that there can still be a variance in the position of the image on the sheet with some digital presses but overall, this issue has been addressed, and printing two sides is not a problem on most digital presses. If the job requires diecutting and binding, be sure to

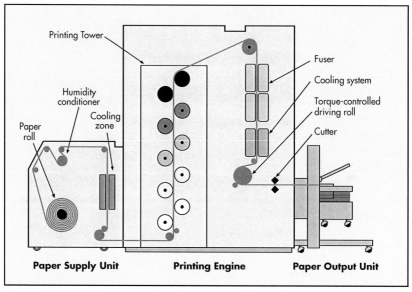

Figure 8-3. The basic principles of operation of the Xeikon print engine.

confirm that the position of the backup is not a concern on the equipment you are considering.

The turnaround time for digital printing is generally shorter than conventional printing and the cost per piece is lower for short runs, but digital prep has helped to balance that out somewhat for convential equipment.

When printing on a digital press, you cannot do much to adjust the color on press. Generally, you have to go back to the file to make any specific color adjustments and then resend the file to the press. It is possible, however, to do a "press proof" prior to running the entire job, and the file can be held in the system pending approval. If it is approved, the job can be run without any additional steps.

Exciting advances in the capabilities of digital presses have enhanced the impact of targeted print materials like customization of documents. Some digital presses print from an imaging cylinder that re-images with each revolution, so you can literally personalize each piece on the press. You are not limited to personalization in just the copy; you can personalize the images as well. Your files must be built to the vendor specifications to utilize this feature, but in most cases, the software is easy to use and your vendor will gladly help to set up your files.

Just like conventional presses, digital presses have imaging limitations, and not all presses have the same capabilities or quality levels. Generally, digital presses use toner, not ink, although some of them do use a liquid toner. Because of the nature of some toners, you cannot assume that the piece can be re-imaged or run back through the press to add colors because the toners can come off. Some digitally printed material cannot be run through a laser printer either since the toners are not laser-proof. Therefore, you should be aware that not all digital printing processes are suited to materials like letterheads or shells.

Currently, most digital presses print only four-color process, but several companies are working on adding a unit to their presses that will allow the use of a fifth color or varnish. If you have a job that is only two-color, you might need to compare the cost of a conventional print job to digital. There may not be a savings since the digital process will run the two-color job as a four-color regardless.

Some of the factors to consider in determining whether or not digital printing is a good option include the following:

- *Critical color.* If it is imperative that the color be an exact match to a product or specific PMS color, digital may not be the best choice.
- *Print quality.* Depending on the type of press being considered, it is a good idea to be aware of the variance in color, the sharpness of the images, etc., that the equipment is generally producing.
- *Quantity.* If the run is more than a couple hundred, you may want to look at the costs of more conventional printing methods.
- *Finished size.* The finished size of the piece can either eliminate the possibility of digital printing or, in some cases, size can make it the only option.

Paper stock, number of colors, finishing requirements, and postpress requirements all need to be considered since there can be limitations in each of these areas depending on the digital press you consider.

The digital press capabilities and limitations vary from one manufacturer to another. As the equipment evolves, the changes address the limitations, expand the capabilities, and improve the quality of the product. Watch these presses—the technology is changing all the time.

Offset Lithographic Printing

Offset lithographic printing (**Figure 8-4**) refers to the process in which an image on a plate mounted on a cylinder is transferred first to a rubber-like blanket and then to the printing substrate. The plates, usually made of metal, are mounted on the cylinder of a printing press. A water-based dampening solution is applied to the surface

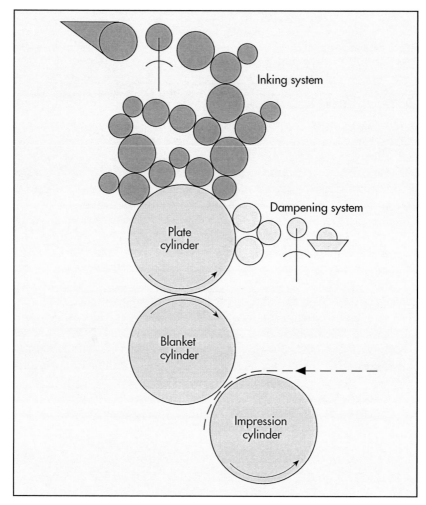

Figure 8-4. The image transfer principle of offset lithographic printing.

of the plate. The dampening solution adheres to the water-receptive non-image areas but is repelled in the greasy image areas. Ink is then applied to the image areas of the plate and transferred onto the blanket from which it is transferred to the substrate (e.g., paper) on which the final piece is being produced. There are variations in this process, and, of course, for the purposes of this text, the process has been stated in very simplistic terms.

Over the years these presses have evolved to accommodate the advances in inks and coatings. They have stretched in length to as many as 16 units, and they have gotten faster, more efficient, and have better controls.

See the discussion later in this chapter on sheetfed and web presses.

Presses with On-Press Plate Imaging

A press with on-press plate imaging, often referred to as a ***direct-imaging,*** or DI, press, is an adaptation of more conventional press technology and incorporates special electronics and platemaking materials. Like the digital presses, there is no need to create film since the digital file information is imaged directly to the press.

Unlike digital presses, the imaging is done to plate material that is pre-mounted on the press, and the image is fixed. These are actual printing plates that are made on the press, not made in the prepress department and then hung on the press.

In addition to saving the time and cost of outputting film and making plates, the makereadies are much quicker because the register of the image is almost automatic. Since the printing is being done from plates using ink, the paper stock limitations, as well as the size and number of colors that the press can print, are determined only by the configuration of the press.

You can adjust the color on press with the computer in the console, much like a conventional press. Also, since the press uses ink, it is not necessarily limited to four-color process. Unlike digital presses, a press proof on a direct-imaging press would require a full set of plates to be imaged and sheets printed to show the color. Unfortunately, the plates cannot be retained or reused once the press images another job, so a set of plates would have to be made for the proof and a second set of plates would need to be made to print the actual run. In a case like this, it is usually less expensive and less time-consuming to have the person needed for approval at the press when it starts to print. If the job is approved, the press can just run out the correct number of sheets and move on to the next job. If changes other than minor color moves need to be made once a job is on the press, the changes would have to be made to the digital file, and then the file would have to be resent to the press and a new set of plates imaged.

Proofing the file prior to direct imaging is usually done with laser printouts or digital proofs. The proofs are not for color but for position and content. When using a digital proof to show color, it is important to know how closely the proofer has been calibrated to the press.

The direct-imaging press is often a good option for jobs that have quantities that are too large for digital printing and have quick turnaround times. The cost to set up a direct-imaging press is generally higher than a digital press, but since the press has good position control, you will have fewer finishing and binding concerns. In addition, there are virtually no stock restrictions, and the inks used can be laser-proof. The number of colors and the image size you can print on a direct-imaging press is determined by the configuration of the press, just like any conventional press. Like a conventional press, you can run PMS or other special colors on these presses.

If price is the only consideration you have regarding the choice between a digital press or a direct-imaging press, the quantity may be the determining factor. While it may be less expensive to run 50 or 100 pieces on a digital press because of the setup cost on the direct-imaging press, you may find that at 500 pieces the unit cost on the direct-imaging press is better. There are no hard-and-fast rules regarding the break point for one method over the other, so when in doubt, bid your job both ways.

Sheetfed Lithographic Offset Printing

Conventional sheetfed presses come in a range of sizes that can print items as small as envelopes to items as large as 77-in.-wide (1,960-mm) press sheets. They can have as few as one printing unit or as many as twelve (**Figure 8-5**). This enormous range of capability allows us to use the equipment that best suits our project. No single vendor has all the various sizes and configurations that are available, nor should they. Printers tend to specialize in certain niche areas, and their equipment reflects those needs. We have printers with as few as two presses specializing in small one- and two-color work and printers with a dozen presses that do long-run four-, six-, and eight-color brochures.

Conventional presses can print on virtually any paper stock within a weight range. The minimum and maximum weights that a press can print varies, but generally a sheetfed press will not run paper that is lighter than 45-lb. book stock nor heavier than 24-pt. board.

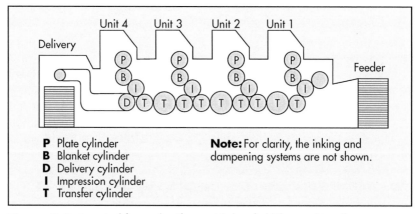

Figure 8-5. A typical four-color (four-unit) sheetfed lithographic offset press.

Most printers have a list of acceptable weights for each press so you can simply ask what the range is if you need that information.

Some presses are equipped with coating towers that put a finish coating on the sheet after the last printing unit. These coatings help to seal the sheet and shorten the drying time. This makes running a sheet back through the press to print the other side possible with less delay and also makes the sheet ready to be handled by the bindery faster. Other advantages to in-line coating include the increased gloss level of the surface; a reduction in the amount of press powder used with the benefit that the sheets are cleaner, and the added protection from scuffing and scratching afforded by the coating. The coatings usually cover the whole sheet, and some cannot be used if the piece will be glued or laminated. Therefore, you should consider all the processes the job will require when deciding whether to use a coating.

Printing presses can be configured to run different kinds of ink. Some presses are configured as **UV presses.** These presses are equipped with special ultraviolet (UV) drying units and print UV-curing inks that set when exposed to UV radiation. A UV press with dryers placed between each printing unit—interstation UV dryers—dries the ink between each color. This results in less ink being absorbed into the sheet on an uncoated sheet and more gloss to the ink on a coated sheet. UV inks appear to be more vibrant because more ink can be applied to the sheet since the lamps make it set up between units. UV presses with interstation dryers are also recommended for printing on plastics, foils, and other nonpaper substrates that do not allow the inks to be absorbed. The dryers set the ink, while conventional inks must dry by oxidation.

Another type of ink that requires a specially configured press is **waterless ink.** These inks are run at controlled temperatures, and the special waterless plates (**Figure 8-6**) have a special silicone coating that rejects the ink in nonimage areas. Waterless printing is not a new concept, but it has not become a big factor in

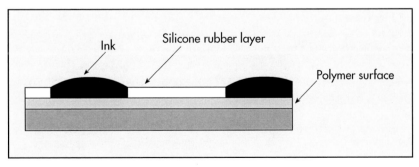

Figure 8-6. Toray waterless plate.

printing in most markets. The special plate used in waterless printing allows the use of high screen rulings, and waterless lithography produces sharper dots than conventional sheetfed printing. The plates are very fragile and more expensive than conventional plates. Should the temperature of the ink rise above acceptable levels, the entire plate can be affected and begin to accept ink in nonimage areas, which means the plate must be removed and remade. The advantages of waterless printing are the fidelity of color achieved from the high line screen capability and the sharpness of the printed dots. The downside is the cost of setting up a waterless system. In addition to the costs of the press with the cooling system and the additional requirements of special plates and platemakers, many printers feel there is not enough value in the quality to offset the additional costs to clients. Waterless presses print on nonpaper substrates better than conventional presses do, and for sharp pristine color they are hard to beat.

Web Offset Printing

A web press prints from a continuous roll of paper stock instead of sheets. Web offset printing is used primarily for high-quantity press-runs like catalogs, direct mail materials, and magazines. Not all paper stock is available in web rolls, and there are maximum weights of paper for most web presses. Generally the heaviest weight that they can run is an 80-lb. book paper or 8-pt. cover stock. It is necessary to check the stock limitations for a specific press.

There are two types of web presses: those with dryers are called *heatset web offset presses,* and those without dryers are called *coldset web offset presses.*

Heatset web offset presses (**Figure 8-7**) are normally used to run coated paper or higher quality uncoated material for projects that require good quality and color. Generally heatset offset webs have four or more printing units and they perfect, or print both sides of the sheet in one pass.

The ribbon of paper is held at a constant tension coming off the roll and passing through the printing units. It then passes through a high-velocity hot-air dryer that removes the solvents from the ink. The web of paper then passes over a series of chill rolls that reduce the temperature of the ink so that it sets before the web is either sheeted or folded on the finishing equipment at the delivery end of the press.

Heatset offset webs, like all presses, have different print areas. In the case of web presses that accommodate a variety of roll widths, the *width* of the roll determines how wide the image can be to print across the web. The web *cutoff,* which is the diameter of the plate

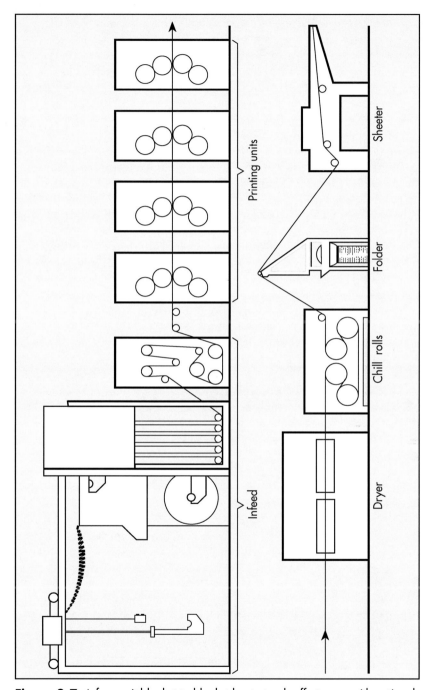

Figure 8-7. A four-unit blanket-to-blanket heatset web offset press with optional delivery to sheeter.

cylinder, determines the size of print area along the length of the web. The combination of the width of the roll and the cutoff is the image area on a web. Roll widths vary, but usually they can be as narrow as 12 in. (305 mm) across or as wide as 40 in. (1,016 mm). Cutoffs are not necessarily the same from press to press even if the same width roll can be run. Always ask your vendor what the maximum and minimum widths are and what the cutoff for a specific press is when trying to plan a web run.

Some heatset webs are equipped with units that make it possible to perform functions like spot UV coating, diecutting, perforation, scratch-offs, and so forth in-line. *In-line* refers to processes that can be accomplished on the press, either as a finishing or folding process, thereby eliminating the need to send the material through another piece of equipment or to a bindery for finishing.

With the right equipment, a job can literally be printed, finished, folded, bound, and trimmed all in-line on the press. For large runs it can mean a great deal of savings in both time and money to have all the processes done in one place at the same time.

The cost of setup for web printing is higher than sheetfed printing, and the press time is more expensive. However, they run faster and can eliminate some, if not all, bindery costs. Therefore, on a per-unit basis, heatset web offset printing is often the most economical way to produce a job.

Coldset web offset presses are generally used to print uncoated offsets and bonds or business reply stocks. These presses are generally used for direct mail materials, letters, continuous forms, books, and manuals. Like the heatset web offset presses, they come in a wide variety of sizes and configurations. However, they differ in that they do not have dryers. Therefore, the substrates have to be absorbent enough to set the ink quickly; otherwise, it will smear in the press or on the finishing line. Depending on the type of work printed on the press, coldset webs can have perforation, folding, and gluing capabilities as well as the ability to create continuous forms for pin-feed applications or friction-fed printers.

On the Press

Once the plates are hung on the press, the *makeready* process begins. Makeready is the series of preparatory steps the press crew takes prior to the actual pressrun. Among other things, the crew makes sure all of the plates are registered to one another and that the images are in the correct position on the sheet. The crew then raises the inking up to the proper levels and balances the color across the sheet. The approved color proofs are used to determine what the client expects the color to look like, and the press crew makes all the adjustments to get the color on the sheet as close as possible to the proofs before the client is shown a press sheet.

Ink is fed onto the plates from an ink fountain that can be controlled by keys that increase or decrease the amount of ink released in a specific zone on each unit; there may be 20–30 such ink key zones across the width of the press. When you adjust the amount of ink controlled by a key, it affects that zone from the top to the bottom of the sheet. In other words, if an image in zone 10, for example, needs more cyan, adjusting the color of that image also affects other images in the same zone. There can be variances in the color from the top to the bottom of the sheet that are caused by limitations of the inking system or the design of the printing form. Talk to the pressroom people to see if the design of the form will cause inking variations, such as mechanical ghosting.

Once the press crew has completed the makeready, the client is brought in for the ***press check.*** The most important thing to remember when doing a press check is to ask questions. These are the pros, and you can learn a lot from them. Guessing or giving instructions that are not well advised can be expensive, if not disastrous.

Several things need to be done during a press check. First, check all of the elements of the piece to be sure that they are in the proper position, are the correct size, and are printing in the proper color of ink. At this time, also confirm that the corrections have been made. Check a rule-out sheet to be sure that all the fold and trim lines are there and that nothing will be cut off or folded. Although all of these checks should have been done at least a couple times by several different people by this time, it is always wise to take one more look. Sometimes corrections are missed or things change from proofs to plates, and you do not want to spend time or waste paper and ink correcting color if the job is going to have to be re-plated.

The second thing that should be done is to confirm that the sheet is in ***register;*** that is, all the images are aligned properly to one another. The easiest way to check for register is by inspecting an area that is reversed to white, such as type. To check register, the area is generally viewed through a 10× (or stronger) ***magnifying glass*** that sits above the printed material so that the dot patterns can be viewed. A number of brands and styles of magnifiers are available; sometimes they are called ***loupes*** or ***linen testers.*** When a hard line such as an edge of reverse type is viewed through a magnifying glass, there should not be a fine ring of any single process color on the edges of the letters or a ragged edge. Register impacts the color, so if the sheet is out of register and you are making color moves, the color adjustments could be a waste of time.

Once the sheet has been checked and is in register, it is time to begin *fine-tuning the color.* If you are using a match color or PMS color, confirm that color with your ink drawdown or PMS swatch. Use the color proofs as a guide for the four-color images, but remember that in most cases the proof is not "ink on paper." Color is very subjective, and there are no absolute rights or wrongs. It is important only that the client likes the printed piece and is happy with the color. When requesting a *color move,* an adjustment in the density or tone of a color, be as precise as possible in explaining what you see and what you want to see in a specific color or area. Listen to the possible ramifications that your requested move may have on other colors or areas before you decide to make a move. Once the color is approved, any marks on the sheet that need to be cleaned up or holes in the print that might be on the plates are marked on the sheet and the sheet is signed. The client should take samples of the approved color; some people wait for the cleanups to be done so their samples are clean.

Not all jobs require a press check, but when they do, it should be done by someone who knows which elements are most important to the client and what compromises will be acceptable if any have to be made. In a case where more than one person will be on the press check, one person should have the final say and sign off on the proof. It is sometimes impossible to give everyone exactly what he or she wants because there are contradictions in the instructions, so someone has to have the last word.

Other Types of Printing

Screen Printing

Screen printing, still known by too many as "silk screening," is a stencil process wherein the image area is open and the non-image area is closed. Ink is moved through the stencil by a resilient squeegee. The stencil/plate is most commonly made from a light-sensitive emulsion, photographically imaged so the areas to print are washed away while the non-image areas are made permanent. The stencil is processed on a fine fabric, most commonly polyester, which holds the parts of the design in place. Without the screen fabric, there would be a problem holding the centers of letters such as D, P, and A in position. Ink is transferred through the open stencil and mesh onto the substrate. **Figure 8-8** shows the key parts of the screen printing system.

Screen printing permits the control of ink film thickness by the diameter of the fibers that form the mesh. It is capable of applying very thick ink films, a strength. It is, however, limited in how thin a film of ink it can produce. It is widely used in the highest quality of packaging labels and in outdoor advertising where qualities including

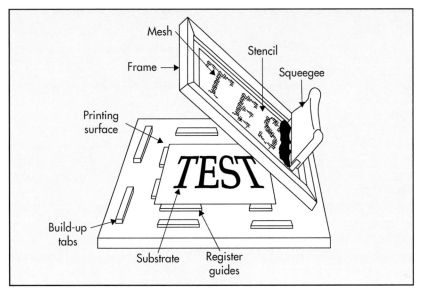

Figure 8-8. The screen printing process.

long life and continuous resistance to the elements are required. Its thick ink film provides more pigment deposit, which facilitates great durability. It is routinely found in both flatbed and rotary applications. It is commonly used in-line with other processes where its thick ink film is desired for visual and/or tactile appeal or for protection, as in applications of clear coatings over other printing.

This process is used for fabric printing like T-shirts and jackets, boards too heavy to run through a press, short-run jobs like signs and posters in limited numbers, as well as printing on plastics.

Flexography

Flexography, or *flexo* as it is often called, is a relief printing process in which the image areas on the printing plate are physically raised above the nonimage areas. Flexography uses flexible polymer plates that are exposed to film and then processed, removing the material around the image so the image is raised. Because the plates are wrapped completely around the small cylinders, the image becomes slightly elongated in the around-the-cylinder direction; to correct for this problem, dimensional compensation is built into the film. In order to proof images, a set of film is output prior to applying the distortion curve; this proof will be used for approval and then as a color guide on press.

Flexographic ink is a liquid instead of a paste-type ink like that used in lithography. The inking system for flexography is also simpler than a litho inking system. In one type of flexo inking system, an anilox roll is

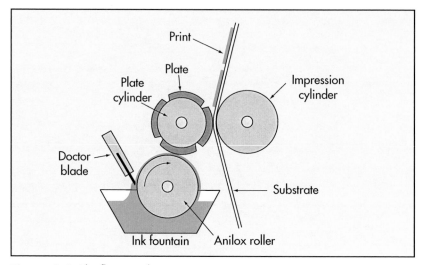

Figure 8-9. The flexographic printing process.

inked, wiped clean (usually with a doctor blade), and transferred to the raised image area of the resilient plate (**Figure 8-9**). The ink remains wet long enough to transfer to the substrate. Because the plate is resilient, made of rubber or photopolymer, it can be impressed against the widest variety of surfaces and print without voids, called *snowflakes.*

Flexo presses, like conventional offset presses, can have virtually any number of units and a variety of finishing options. Flexo presses can print static-cling material, plastic, foils, as well as conventional paper stock. The finishing processes that can be done in-line include coating, diecutting, sheeting, and perforating. Some flexo presses can run multiple webs of stock in-line and marry them on the end of the press creating layered products like instant response coupons (IRC) and peel-and-reveal game pieces. Flexo-printed images are not as sharp as offset, and the image trapping lines are thicker, but these concerns are often offset by the cost-efficiencies and the flexibility of these presses.

Gravure Printing

Gravure printing, an intaglio printing process, employs a precision metal cylinder with image areas composed of tiny cells engraved or etched into the metal surface. This cylinder is simply rotated in a pan of ink, and its nonimage surface is wiped (doctored) with a very thin metal, plastic, or other synthetic composition blade called a ***doctor blade.*** The doctor blade wipes the smooth surface clean of ink, leaving the image areas (cells) filled with ink which transfers when impressed against the substrate.

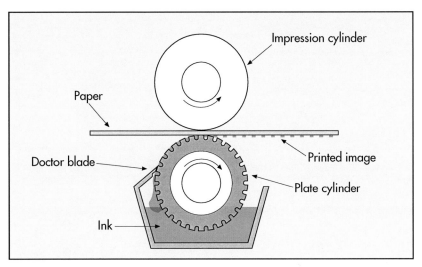

Figure 8-10. The gravure printing process.

Figure 8-10 illustrates the gravure image carrier. There is only one moving part. Since it is a "direct" process, meaning the imaged plate prints directly to the substrate, it requires a smooth receiving surface that contacts all the cells or there will be voids in the image, sometimes described as snowflakes.

Gravure is used for publications, catalogs, Sunday newspaper supplements, labels, folding cartons, flexible packaging, gift wrap, wall coverings, floor coverings, and a wide variety of coating applications.

9
Finishing

Diecutting

Diecutting is the method used to cut finished pieces to shapes other than square. Diecutting requires a *cutting die,* which is a board that has channels cut in the shape of the piece, with channels for scores, perfs, and holes as needed. Then the appropriate metal strip is inserted into each channel for the function required. Sharpened metal strips or *"knifes"* are bent to the shape of the cut edges and secured in the board with the sharp edge up. *Scores* are round-top metal strips that are inserted in the board where the folds in the piece are required. *Punches,* sharpened metal cylinders, are inserted in the board to make holes. *Perfs* are metal strips with sharp raised areas— teeth—that cut through the material to form a perforation. The weight of the material and the strength of the piece determine how close together the teeth are and the width of each tooth.

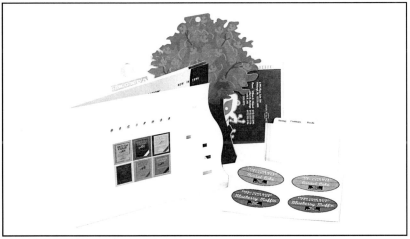

Figure 9-1. A variety of diecut products. (From *Binding, Finishing, and Mailing: The Final Word* by T. J. Tedesco, Printing Industries Press.)

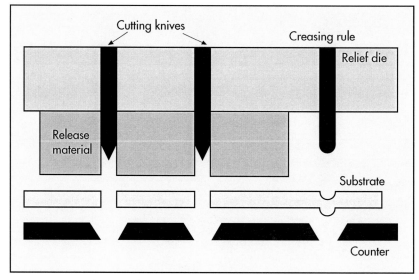

Figure 9-2. Side view of a die board (top) and counter (bottom).

A die can have one or all of these elements depending on the design. The size and complexity of the die determines the cost. Dies can be made one of several ways. ***Handcut dies*** are made by mounting a ***die line***—drawing of the piece layout done on frosted acetate—on a piece of wood and then cutting the board with a jigsaw using the layout as a guide. The metal is bent to the required shape by hand and inserted in the wood.

Laser dies are a more advanced method. The layout is entered into a specially designed computer system that generates the die line. A laser then cuts a blank sample from the specified stock to make sure all the scores are in the proper position, the perfs are strong enough, and so on. Once the samples have been approved, the die board is cut with a laser from the same file. A separate piece of equipment bends the metal using the same information. The bent knifes are manually inserted into the channels in the board. Very intricate or sharply angled designs may require special metal rules that are not available from every supplier. This can add both time and cost to the die.

Another method of diecutting that can produce very intricate patterns is gaining popularity. ***Laser diecutting*** uses a laser beam to burn away the paper to create the opened areas. This method is used on stationery, cards, and covers, primarily for internal design cuts, but not perimeter shapes.

Pockets and Flaps

When specifying a folded element to hold loose material either in a book or folder, you need to be sure you are specifying the correct configurations. A *pocket* has a tab that glues to the body of the piece on one or both sides and keeps material from falling out (**Figure 9-3**). A *flap* is generally just folded in to form a resting place for materials or a place to diecut slits for a business card that will not show on the outside. A flap will not secure the material in it like a pocket, as the sides are open (**Figure 9-4**).

There are many other custom features that are done on diecutting equipment such as business card slits, slots for CDs or DVDs, and slits in the face of a full pocket to insert a sheet. Generally, anything that is not a square-cut folded piece requires a die to get the flat shape that can be folded into the final product.

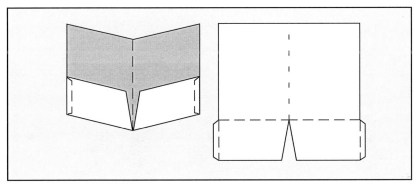

Figure 9-3. A diecut folder that has two pockets with glue tabs.

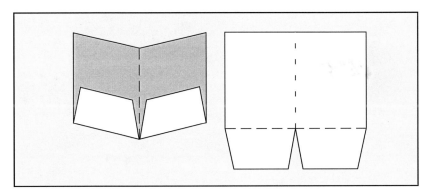

Figure 9-4. A diecut folder with two flaps.

Foil Stamping

Foil stamping uses metal plates called *foil dies* that are mounted on a diecutting press that has a heating unit to impress special *foil* onto the substrate. The foil is fed into the machine between the die and the paper. Under precisepressure and temperature settings, it is released from its backing material and adheres to the paper in the pattern or design of the die.

Foil is available in a wide range of colors, finishes, and materials. The most common types of foil are metallic. They are used on Christmas cards and corporate materials usually in gold or silver, but there is a wide variety of metallic colors, tones, and finishes available. Foils are also available in nonmetallic materials that range from matte to high gloss and from clear to primary colors. Foils can be used as subtle tints or as bright bold design elements. Specialty foils are used for *scratch-offs* and *write-in* areas on materials like game cards and credit cards.

These various foils have different heat and pressure requirements, and the design and the paper have to be considered when choosing a foil. Some foils are not suitable for very intricate designs or sharp

Figure 9-5. Examples of foil-stamped products. (From *Binding, Finishing, and Mailing: The Final Word* by T. J. Tedesco, Printing Industries Press.)

angles because of the release, some are not suitable for large areas, and some require so much pressure and/or heat that they will blister the stock. It is important to confer with the vendors about the foil best suited to the design and paper you are considering. Catalogs available from the finisher can help you to determine the color and finish of the foil you want, but the vendor will have to order the foil that is right for the job. Whenever possible, it is a good idea to proof the die with the foil on the exact paper you will be using to confirm that the die is right, that the foil and paper are compatible, and that the client likes it prior to running the job.

Stamping dies are made of various metals. The length of the run, the type of material, and the intricacies of the design are all factors in choosing the type of metal to use. These factors, as well as the size of the die, determine the cost.

Foil is purchased in rolls in the width required to cover the image; the length is determined by the size of the press. The number of square inches of foil required to cover the area determines the cost. The area around or in between foil-stamped elements is wasted foil and, in many cases, cannot be reclaimed. Some presses can run more

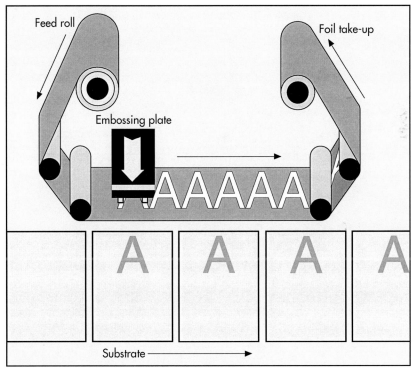

Figure 9-6. Typical hot foil-stamping configuration. (Courtesy Bobst)

than one foil in the same pass; rolls in the proper width for each element are loaded on the feeder and fed through the press to align with the design element on the die and the sheet.

Embossing

Embossing dies are made from metal dies that create raised areas in a specific design in the paper, sometimes on several levels. *Debossing* is the same process except that the design or element is pressed down into the paper. The dies have a counter element that forces the paper under pressure into the die, and that area of the paper is raised or depressed forming the design in the paper. When embossing is done in register with a printed element—for instance when a printed logo on the sheet is to be raised—it is called *register embossing.* When an embossed element is applied to an unprinted sheet or not positioned exactly to a printed element on the sheet, it is called *blind embossing.*

Embossing dies, like foil dies, are made in various grades of metals. The same factors are considered to determine the appropriate metal for your design. The cost is determined by the metal size and complexity of the die. Heat is not generally required in the embossing process, but it can be used to enhance the effect on some papers.

Foil emboss is just what it sounds like—a combination die that embosses and foil stamps an element in a single pass through the press.

Figure 9-7. A variety of products that have been embossed. (*From Binding, Finishing, and Mailing: The Final Word* by T. J. Tedesco, Printing Industries Press.)

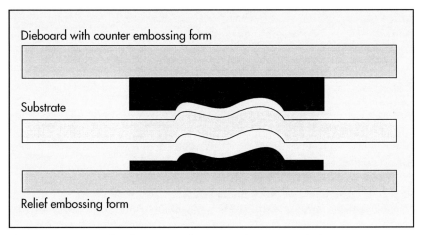

Dieboard with counter embossing form

Substrate

Relief embossing form

Figure 9-8. Typical embossing configuration. (Courtesy Bobst)

Lamination

There is often a great deal of confusion about the term *lamination,* since there are two completely different processes that are both a type of lamination. All the word "lamination" tells you is that something is going to be affixed to something else. Therefore, it is important to specify if you are asking for two sheets to be pasted together or if you want a clear Mylar or plastic material applied to the printed piece. The two processes are not necessarily mutually exclusive, and the costs associated with each can be radically different, so be specific when referring to lamination to be sure all involved parties are talking about the same thing.

Film lamination is the application of a clear material—usually a Mylar or polypropylene byproduct—to one or both sides of a printed piece. Film laminates are made from several different types of material and come in a variety of weights. The best material and weight for your application will usually be determined by the way the piece will be used, the environment in which it will be used, and the life expectancy of the piece. The primary reason for applying a film laminate to a piece is to provide protection, so ask the vendor for recommendations based on the projected use of the piece.

Some pieces are film-laminated on one or both sides and then trimmed or diecut to the finished shape; these are *clean-edge* or *flush laminations.* Other pieces have what is known as a *sealed edge* or are *encapsulated:* these pieces are trimmed to final size and shape and then laminated on both sides with film material that is larger than the printed piece. The finished piece will have a "frame" of clear plastic sealed on the edges around it. The frame can be whatever width you require and can vary from one side of the piece to another. Luggage tags, for example, often have a wider frame of sealed material on the top to allow for a hole that the chain goes through. The weight of the material as well the number of square inches used on each piece determines the cost.

If a sealed-edge piece is to be diecut into a shape with clear edges, you would need to have it diecut and then laminated on two sides with enough laminate around the piece to mount a second die (made to the same shape but larger by the width of the frame required), and finally diecut again.

Sheet-to-sheet lamination is the process of gluing two sheets together on a piece of equipment known as a ***laminator*** or ***paster.*** Sheet-to-sheet lamination affords a number of options. For example, you can glue two printed sheets together to form a two-sided piece of heavier weight, you can glue a printed sheet to chipboard or corrugated board to make a display, or you can glue a printed sheet to either side of a piece of chipboard or corrugated board to create a two-sided header or display.

A few factors need to be considered when planning a job with sheet-to-sheet lamination. The weight of the material may cause a curl in the finished piece. Sometimes when the job calls for one printed sheet to be laminated to a relatively thin sheet of chipboard, the vendor may recommend backlining or laminating a sheet of blank white paper to the other side to offset the curl and make the finished piece more attractive. The type of equipment being used determines whether the lightweight sheet or the board is the element that the glue is applied to. Usually the only concern for the client is that the finished product be straight, flat, and, in the case of two-sided print material, that the alignment from one side to the other has a minimum variance. Not all sheet lamination equipment can guarantee the same precision in placing the sheets, so if your job has critical alignment issues be sure that is understood from the start.

10
Binding

In this section, we will discuss some of the common bindery methods, but keep in mind that these represent only the basics. Many other options and variations are available for the jobs that just don't "fit" conventional binding methods. Talk to your vendors and let them make recommendations on ways to approach bindery issues. More than once in my career, I have taken a project to a bindery where conventional binding just would not work, and they have "invented" a process or piece of equipment to address that specific set of binding challenges or simply suggested another way to approach the layout that would allow for less complex binding operations. It can't always be done—or sometimes the fix is more expensive than the project budget will allow—but you never know until you ask. Another reason for involving the bindery early on in job planning is that you do not want to get into production only to find you cannot bind the job or the time requirements have been miscalculated.

Folding

Single-Sheet Materials

Let's start with folding combinations. When the piece is one sheet of paper that is folded to the final size, there are a number of terms used to describe the exact direction or combination of folds you are using.

A *French fold* is a piece that is folded down from the top in half and then over from left to right in half. This used to be the common greeting card fold because the paper used could be a lighter weight and all the printing was done on one side (**Figure 10-1**).

A *roll fold* is when each panel is folded over the panel to its right (or left); the last panel forms the "cover." Each panel starting from the left will be a slightly smaller size than the panel to the right so the folds will nest into one another and the piece will lay flat. You will need a blank comp to finished size on the weight and type of stock to be used to determine the exact amount each panel needs to be reduced (**Figure 10-2**).

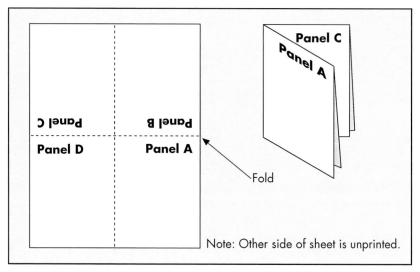

Figure 10-1. A French fold, a printed piece that is folded down from the top in half and then over from left to right in half.

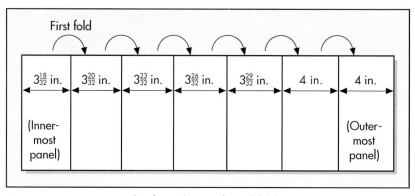

Figure 10-2. An example of panel sizing for roll folds.

A **single fold** or **four-panel fold** is just what it implies—a single sheet of paper folded once to form a four-page signature. If you will not be folding the sheet in half, you will need to specify the width of the short panel for proper placement of the score (**Figure 10-3** and **Figure 10-4**).

A **letter fold** or a **six-panel fold** is a sheet that is folded in two places. When the printed piece is in a vertical format, it is referred to as a six-panel brochure. The right panel is folded in, and the left panel is folded over. The right panel must be slightly smaller in width than the center panel, while the left panel is slightly wider to line up

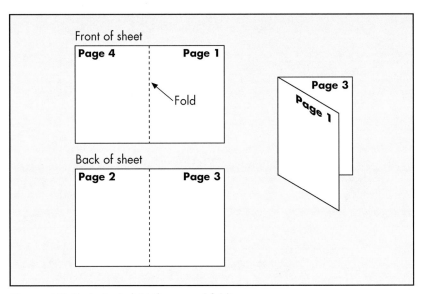

Figure 10-3. A single fold (four-panel fold).

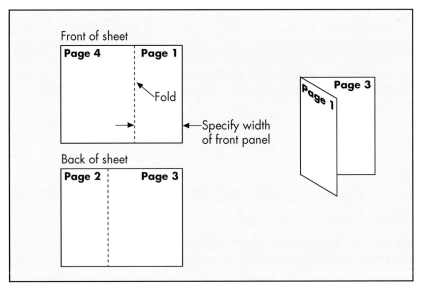

Figure 10-4. A single fold with a short front panel.

with the right edge fold. When in a horizontal format, it is referred to as letter style. The bottom panel folds up, and the top panel folds down. The bottom panel is shorter than the center, and the top panel is slightly longer. (See **Figure 10-5** and **Figure 10-6.**)

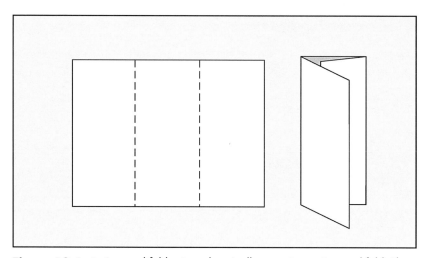

Figure 10-5. A six-panel fold oriented vertically, creating a six-panel fold. The panel that is folded inside the other two should be slightly narrower.

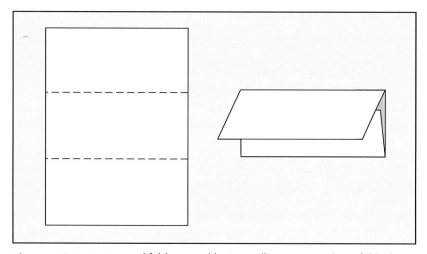

Figure 10-6. A six-panel fold oriented horizontally, creating a letter fold. The panel that is folded inside the other two should be slightly narrower.

An *eight-panel/double-parallel fold* is when a flat sheet of paper is folded to half its width and then in half again to one quarter of its original width. The first fold can result in uneven open edges so you should be sure that the stock you have chosen will fold over itself smoothly and lay flat (**Figure 10-7**).

An *eight-panel/double-gatefold* is a sheet that has three folds and the left and right panels are folded into the center and then the

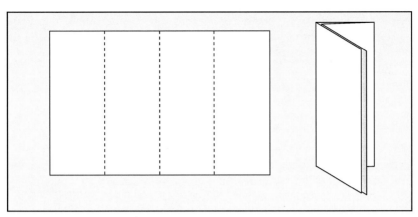

Figure 10-7. An eight-panel/double-parallel fold.

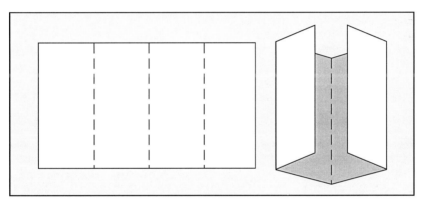

Figure 10-8. An eight-panel/double-gatefold.

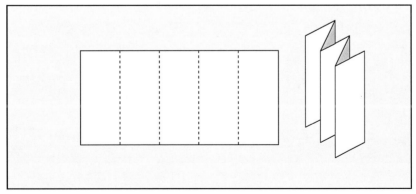

Figure 10-9. An accordian fold, or Z-fold.

piece is folded left over right. You will recall reading about gatefolds as they impact page count; this is the same fold configuration used as a finished piece (**Figure 10-8**).

With a **Z-fold** or **accordion fold,** each panel is the exact same size and they are folded behind one another in alternating directions to form a Z or an accordion when viewed from the top (**Figure 10-9**).

Folded Press Signatures

When a sheetfed printed job is delivered to the bindery, it is usually in the form of flat press sheets. For multi-page books, the press sheets are laid out so that, when they are folded into signatures for binding, the pages will be in the proper sequence (**Figure 10-10**). This means that the pages that were printed on that press sheet are never cut apart, do not need to be individually collated, and cannot get out of order with one another. For instance if you are producing a 16-page 8½×11-in. book, the press sheet is laid out so that, once printed, it can be folded much like a French fold down to 17×11 and then once again to 8½×11. The pages are now in proper sequence, and the signature is bound either as is or married to a cover. For books with larger page counts, multiple folded signatures are laid out to fold and nest into (as with saddle-stitched publications) or stack upon (as with perfect-bound and case-bound books) one another for binding in order. The number of pages printed on a press form is determined by the size of the press and the specifics of the job, such as some of the following considerations:

- If color is critical, it may be easier to control the color on smaller forms.
- Gatefolds that bind in cannot be folded into signatures.
- The page count of the publication is not an even number of sixteen-page signatures, but is in even "eights."
- Run length.

This is part of the planning that a printer does prior to quoting a job. If you are to furnish mechanicals in press form layouts, the printer can supply you with the layout grid for each press form for each job.

Saddle Stitching

Saddle-stitch binding, probably the most popular form of combining signatures, is widely used for everything from magazines and catalogs to church bulletins and programs for public events. In this process, one or more signatures are stacked on top of an angled metal surface called a **saddle.** The folds in the signatures are carefully centered on top of the saddle, and then a wire is stitched through the stack of

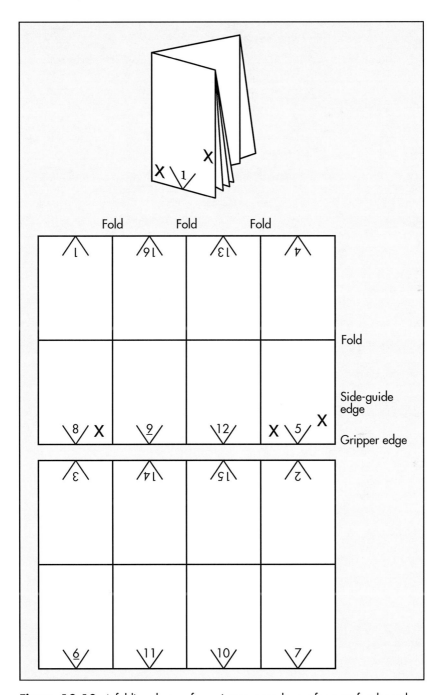

Figure 10-10. A folding dummy for a sixteen-page layout for a perfect-bound book (top) and the dummy opened up showing imposition for front and back forms.

Figure 10-11. A press sheet entering the first buckle on a buckle folder.

paper to bind the signatures together (**Figure 10-11, Figure 10-12, and Figure 10-13**). These wires are aligned in the same direction as the fold and should pierce the fold itself. Saddle-stitch binding can be done in small quantities with a modified stapler, using standard office-variety staples. Most saddle-stitch binding, however, is performed with an automated binding machine using a conveyor chain and a high-speed stitching head. Like a massive needle and thread, this high-speed method uses rolls of thin wire that are cut, bent, and inserted as needed. When staples are created from a roll of wire (as opposed to being pre-manufactured), the process is called *stitching* instead of stapling.

There are usually two staples or wires centered an equal distance from either edge of the finished piece. On oversized books or books with high page counts, three or more staples or wires can be used.

This method is used on books of fairly high page count depending on the weight of the stock used in them. Bindery professionals and blank comps can help you determine when a book is too large to saddle-stitch. There is no minimum sheet count or finished size for stitching. Two or more sheets of any dimensions can be bound by this method.

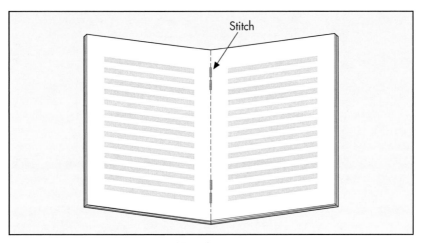

Figure 10-12. A saddle-stitched booklet.

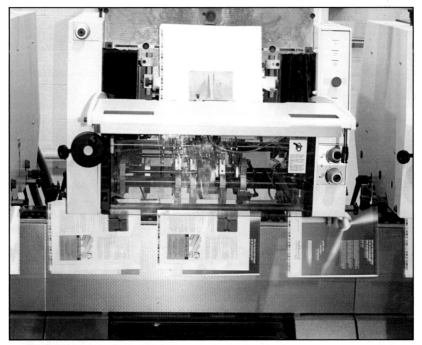

Figure 10-13. A saddle stitcher.

When you have more than one type of material being bound into a common cover, the order of the sheets determines how the sections are bound in and the number of wires that will appear on the spine. For example, suppose a booklet has a 16-page signature printed on gloss book stock, a 16-page signature printed on safety paper, and a 4-page cover printed on text stock. Each 16-page signature would be bound separately into the cover with staples or wire. Usually two wires or staples are used for each signature, and the spacing is adjusted to make both sections secure. So, in this example, there would be four wires or staples on the spine of the finished piece.

Normally in saddle stitching, each signature in a publication is opened to its center and dropped on top of the previous signature already on the saddle of the saddle stitcher. One variation involves saddle-stitching one signature, trimming the signature on three sides, inserting a second signature into the first signature (but not the middle of it), and then saddle-stitching the two signatures to the cover. This procedure produces what is referred to as a **wrap fly hidden stitch.** Using a publication with a 16-page signature printed on gloss book stock, a 16-page signature printed on safety paper, and a 4-page cover printed on text stock as an example should make it clearer. In our example, the sixteen-page signature printed on safety paper would be saddle-stitched and trimmed on three sides. Next, this "bound" signature would be opened after page 4 and the sixteen-page signature printed on gloss book stock would be inserted there. Next, both signatures and the cover would be saddle-stitched through the center of the signature printed on gloss book stock. In this particular configuration, the publication would consist of the front cover then four pages of safety paper, sixteen pages of the coated stock, twelve pages of the safety paper, and the back cover. Since this variation requires additional steps, consult with the binding vendor regarding feasibility and cost.

When a book is planned for this binding method, the stock must be chosen with the bindery in mind, and blank comps are a must because the stock that is bound first must be strong enough to support the weight of the second signature and the cover must be heavy enough not to break under the uneven pull of the sheets, allowing the material to pull through. This method requires more handwork and multiple bindery steps, so it is more expensive than conventional binding. However, the finished piece will have a smoother spine with only two wires showing, so in some cases it is worth the additional time and expense.

Perfect Binding

Perfect binding, a form of **adhesive binding** that uses glue to hold pages together, is an especially versatile binding method. Books as thick as 2⅝ in. (66.7 mm) can be perfect-bound. Signatures and single leaves are assembled into a book block by stacking one component upon another. The backbone is then roughened to expose the paper fiber, which is a better surface to hold the adhesive. Hot-melt glue is applied to these edges, the cover is wrapped around the book block,

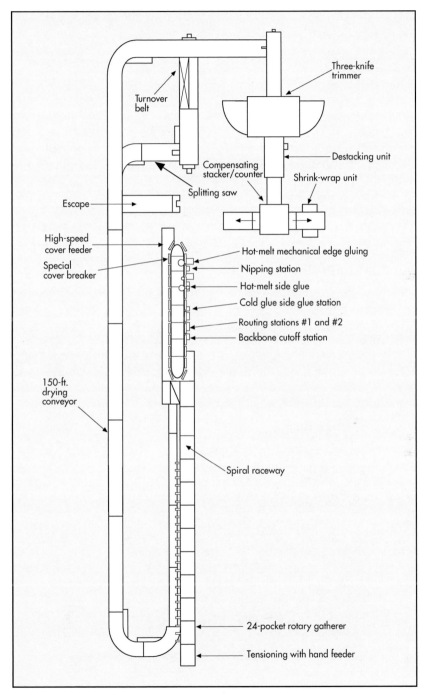

Figure 10-14. Perfect binding line. (Courtesy of Muller Martini, Inc.)

and it proceeds through a three-knife cutter to trim the head, foot, and face. **Figure 10-14** shows a ***perfect binding line.***

While perfect binding is more expensive than saddle stitching, there are fewer restrictions on sequences of multiple stocks. For example, pages can be calculated in twos, not fours. In addition, you can bind in tabs or pull-outs, and the finished piece has a more elegant appearance. This method is not effective if the book has too narrow a spine as you can not get a clean box score on the spine of the cover, and if the surface area for gluing is not large enough, the book may tend to come apart. A general rule is a spine of ⅛ in. (3 mm) or more.

In some cases, the cover has an additional score on the front and back cover and is glued to the front and back sheet to create a hinge. This reduces the stress on the spine and adds a cleaner finish to the look of the inside front and back cover spreads.

Plastic-Comb Binding

Plastic-comb binding is used frequently to bind presentations and workbooks, among other products. Many people refer to plastic-comb binding as GBC, which stands for General Binding Corporation, a major manufacturer of the plastic combs (**Figure 10-15**).

In plastic-comb binding, the sheets are punched with rectangular holes on the binding edge with a special piece of equipment. The comb is then placed in the binding equipment to be opened. The punched pages are then inserted, and the comb is closed.

These combs can be opened and pages removed or replaced and then the comb re-closed. The piece will lay flat, but the spine will not allow the pages to be turned back past the width of the spine. The plastic combs come in various sizes and colors and can be customized

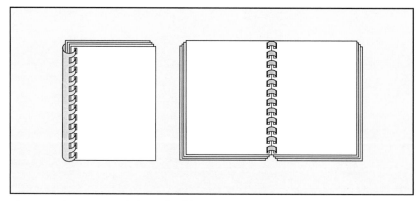

Figure 10-15. A comb-bound publication.

for a project. For instance, you can order spines imprinted with the client's name or book title. Since this is both inexpensive and versatile, many companies own this type equipment.

Wire Binding

Wire-o and *Spiral* are two types of wire binding. Both require a punch on the binding edge and allow the book to lie flat like GBC binding. With either Wire-o or Spiral binding, the sheets can be flipped to the back like a flip chart, but they cannot be opened and updated or reused. The wire configuration and equipment required for each of these methods is quite different, as is the finished appearance.

Spiral is a continuous coil of wire, and the punch produces a small round hole. This type of wire binding is used extensively on school notebooks. The maximum size of a Spiral spine is 30 mm. Because it is a continuous spine, the rigidity of the material—especially the cover—determines how stable the book will be. Spiral is available in both standard wire colors as well as plastic-coated wire in a myriad of stock colors. For large orders, a custom color can be created. Ask your vendor for minimum quantities and lead time requirements.

Wire-o is a continuous coil of preformed wire that has a loop configuration. The punched hole is rectangular, and the formed wire is put through the punched sheets and then crimped. Because of the looped construction, there is more rigidity to a Wire-o spine than with Spiral

Figure 10-16. A variety of mechanical-bound publications. (From *Binding, Finishing, and Mailing: The Final Word*, by T. J. Tedesco, Printing Industries Press.)

in larger diameters. Wire-o is available in a wide range of sizes up to 28 mm and a number of colors, both bare wire and painted. Wire-o can be custom-ordered; request minimum quantities and lead times from your vendor.

Planning for the Bindery

Every binding method has an impact on how the materials need to be printed. Some methods are more forgiving than others, but it is always wise to review the bindery before the materials are prepared to print.

Any publication that is to be **saddle-stitched** or **perfect-bound** must be printed in **printer's spreads** to conform to the finished page order and sequence in the bound book. With either binding method, the press sheet would be set up so that the first and last pages of a signature print side by side on the same side of the sheet, with the second page and next to last page side by side on the other side, and so on. (See **Figure 10-10** earlier in this chapter for a sixteen-page signature for a perfect-bound publication.)

When you are using multiple stocks in conventional or alternate saddle-stitch bindings, the printer's spreads may not be that straightforward since your stocks may be mixed and function as separate books for layout. In these cases, it is almost imperative that the book be built into a blank comp and the page numbers and content noted, then the marked-up blank comp can be used to determine how the pages need to be laid out for press.

In order to confirm the page sequence, it is a good idea to request a blank comp in the specified stock for any multipage book regardless of the stocks and binding configuration. In stitched or perfect-bound materials, there are issues like the actual live area of each page that are determined by the impact of binding, the number of pages, and the weight of the stock.

Comb- and wire-bound books are laid out as single sheets, not spreads. If the book prints on both sides, then page 1 is backed by page 2, and so on. Blank comps or dummies of any of these binding types in the specified paper are a good tool for determining the live area to ensure that none of your copy will be in the punching or binding area of the book, as well as the overall weight of the publication.

If final files are to be furnished by the client or agency in printer's spreads, furnish them with the blank comps and review with the client the layout and live area issues before they build the print-ready files.

Regardless of who built the printer's spreads, always output digital or laser proofs and bind them to confirm that the sequence is correct and that all the elements are within the live area before any film or plate work is performed. This may take a day, but it could save a week's time and a lot of money later.

Additional Bindery Processes

The size and configuration of the piece determine whether folding and gluing can be automated and done on a machine or must be done by hand. Manual operations are used for very small orders, oversize pieces, and pieces that have glue patterns or folds that the automated equipment cannot do.

The automated folding process is done on machines that use a series of belts and rollers that are set to create right-angle folds at precise intervals. These machines can fold the sheet and then turn it and fold it again if that is what is required. If this piece does not stitch or bind, it is usually finished at this point. (See **Figure 10-10.**)

Some folders feed directly into the stitcher. In these cases, if the printed piece is a saddle-stitched book, the piece can be folded and bound in one operation. If the folder does not have a stitcher on it, the folded piece is moved to a separate machine for stitching. Some folder/stitchers have a trimming unit on the delivery end, called a ***three-knife trimmer.*** This unit does what its name implies—it cleanly trims the piece to its finished size. If there is no trimming unit on the folder, the piece is finish-trimmed on a cutter and is then complete.

Automated folder-gluers are used for pocket folders, video cases, and other materials that have straight-line glue patterns. After the piece is diecut and the excess material has been removed leaving only the finished flat shape, the piece is fed into the machine flat. An arm folds the glue tab in, and the glue is applied. The right-angle fold is made, affixing the tab to the inside of the body of the piece. The glue used in this process is usually hot-melt glue that sets up very quickly. If an additional fold is required to finish the piece like closing a pocket folder, a second right-angle fold is made and the piece is ready to insert or pack.

Collating and ***inserting*** are the steps taken to put multiple elements in order into a carrier or folder. This can be done by machine or by hand depending on the elements to be gathered and the number of units to be assembled. Usually the best way to handle collation is to have all the elements of the piece or section finished and in the bindery prior to beginning the collation process since the fewer times you have to handle a piece the less expensive it is. When this is not possible, partial collations can be done to save time when the last pieces are available. Insertion is simply putting the materials into a pocket folder or box to complete the piece.

11
Packaging

Packaging a product for shipping or delivery and the cost associated with packaging are often overlooked when building the budget. Various factors enter into the packaging, such as the number of shipping addresses, product life span, and the amount of protection required to have a piece arrive at its final destination in good shape.

There are a variety of ways to package printed materials. Following are a few common methods.

Paper Wrappers

With a paper wrapper, the product is wrapped in kraft paper and sealed with tape. This can be done to any size or weight of material. In the case of very small quantities or lightweight materials, it is sometimes recommended that a stiffener, usually a piece of chipboard, be used to make the package more rigid.

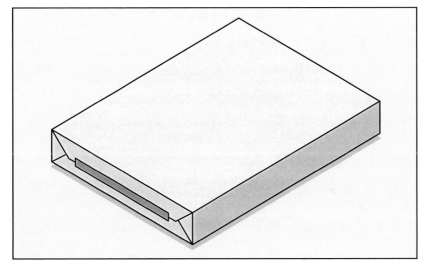

Figure 11-1. Product covered using a paper wrapper.

Since this kraft paper is opaque, you would need to consider putting some form of identification on the package. This identification could be a printed end label or a stamp that identifies the material and the count, or a copy of the material could be taped to the outside of the package. The material costs are not as expensive as with shrink wrapping, but the kraft paper is not as protective if the product is exposed to moisture.

Shrink Wrapping

Shrink wrap is a tubular-shaped, heat-reactive plastic supplied on a roll. The product to be wrapped is placed inside this plastic "tube." Next the material is cut by a heated knife creating a four-sided loose "bag" around the product. When exposed to heat, the plastic "draws up" and conforms to the size and shape of the materials. In application, a conveyor belt carries the materials through a tunnel that houses the heat source. The speed of the belt and the temperature are adjustable and have to be set to allow the shrink-wrap material to draw up properly on each product. Shrink wrapping small quantities

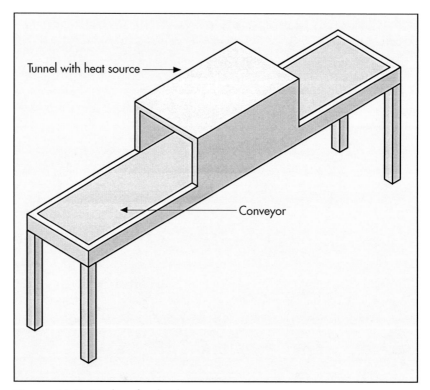

Figure 11-2. Machine for shrink wrapping.

or lightweight material often requires a stiffener, usually a piece of cardboard on the bottom or on both sides to keep the materials flat.

Shrink wrap is available in a variety of thicknesses. The size, weight of the package, and various other considerations determine the appropriate thickness to use. The size of the tunnel limits the size of the materials that can be shrink wrapped in it. The cost per package is determined by the thickness of the shrink wrap, the amount of shrink wrap used, and the cost of any stiffeners required. This number is multiplied by the number of packages to be produced.

Banding

Banding is the process of securing a stack of material with either rubber bands or strips of paper. The bands are wrapped around the middle of a stack of materials to secure them in a specific quantity. Sometimes two bands are used—one lengthwise and one widthwise around the material—to prevent the materials from slipping out in transit. Using bands is less expensive than full wraps, but the material is not as secure and bands may cause some loss of product.

Bulk Packed

Bulk packing refers to stacking unwrapped material in a box of the appropriate size. The box size is determined by the size of the piece and the pack-out quantity requirements. Weight and size requirements and shipping method determine the strength of the material from which the box needs to be made.

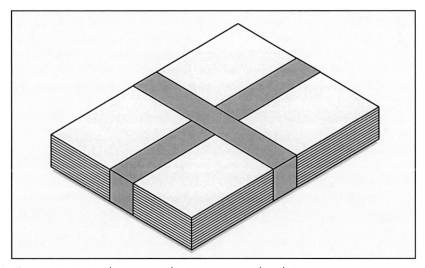

Figure 11-3. Product wrapped using two paper bands in a crisscross pattern to minimize slippage of material during shipping.

Box-packed materials are often repackaged in larger cartons. In some instances, the boxes can be made from chipboard. However, if the boxes are being used as shippers or stacked on skids, they must be made from corrugated board strong enough to resist damage during shipping.

If materials are to be stacked in the box side by side or in layers, dividers of cardboard or corrugated board can be used to prevent product from shifting and becoming damaged.

Carton Packed

Carton packing indicates that packaged material has been placed in a box of the appropriate size for shipment. Like the bulk-packed boxes, the size and strength are determined by the product size, weight, and shipping method.

Both bulk-packed and carton-packed projects sometimes require that a custom-size box be ordered from a box company or made on a box cutting machine at the vendor. If the product does not fit in a standard box snugly, there may be too much air or empty space when the box is sealed. To fill the space, additional packing material is added to keep the product from shifting and to lessen the possibility that the box will be crushed if something heavy is placed on top of it. A custom box may add a little to the cost of the boxes, but it is usually minimal. Using additional packing material with a standard box can be more costly than the custom box charge and not as effective as using a properly sized box in preventing product damage.

Skid Packed

Skid packing refers to stacking any material on a wooden or corrugated platform that can be picked up and moved by a forklift for shipping or storage. There is a skid charge for each skid in addition to the packaging charges. Very often *stretch wrap*—a strong flexible plastic material that when drawn tightly around an object conforms to the shape and adheres to itself—is wound around the packed materials on a skid to hold it in place as well as to add a layer of protection from the elements; this is called *skid wrapped.*

When working with promotional materials and some direct mail projects, it is sometimes necessary to *kit* the materials. This process requires that a carton be found or made that will accommodate a specific number of each of the elements of the promotion. Kits should provide maximum protection of the materials, should be easy to pack, and should be economical to ship.

Flat Corrugated Shippers

Flat corrugated shippers are sometimes used to ship headers or posters. These shippers are basically large sheets of corrugated board folded in half and secured on all open edges with tape or staples. If your material is relatively flat and oversized, this is often a better alternative than folding the piece to fit in a box or envelope and can be very economical. Some displays have been designed to fold the print area inside and form self-shippers.

Packaging Considerations

Each additional type of packaging your job requires has costs associated with it as well as time requirements to complete. For product that must be shipped over long distances or delivered by general package delivery companies, the packaging can be the difference between having a usable product delivered or not.

It is always preferable to plan for the packaging of your job when you are submitting the specifications so the costs can be included in the original estimate. This ensures that they will not be overlooked and come as a complete surprise on the final billing. If the vendor knows the packaging requirements when he starts the job, all the necessary materials can be secured while the job is being printed, helping to avoid delays at the end of the job.

If you cannot determine the packaging requirements until after the job has been started, be sure to get the information to the vendor as soon as you can and advise the person approving the budget that those costs will be forwarded to them as soon as you have the information on the packaging requirements.

Do not just assume everyone knows you have to pay for packaging; make a point of the fact that it is not included. It is often tempting to scrimp on the packaging, but this can be shortsighted if you jeopardize the condition of the material in shipping. Ask your vendor to help you determine the extent of the packaging required to ensure that the product arrives in as pristine a condition as it is when it's shipped.

12
Shipping

Freight and shipping charges vary widely depending on a number of factors, including weight, size, shipping destination, and arrival date and time. When the project has to meet a specific delivery date, the choice of freight services available to make the deadline may be limited.

It is always best to consider the costs associated with the various freight options as early in the process as possible, so if time becomes an issue, you know what the most cost-efficient options are. Knowing the freight costs in advance can help to determine whether it would be less expensive to pay for overtime in a print shop or bindery than to have to send the material next-day air freight to make a deadline.

Without the specific addresses and quantities to be shipped to each address, it is impossible to develop an exact freight quote. There are so many options and services to choose from that it would be like asking for a quote on airfare if you did not know when you wanted to leave or where you wanted to go. However, using general quantities and a list of the cities to which the material is going to ship, it is possible to develop a working estimate based on the time available for delivery and the weight of the piece.

U.S. Postal Service

The U.S. Postal Service (USPS) is the most common method of sending catalog materials, direct mail, and business mail. The postal service offers many services ranging from registered mail to parcel post with delivery options that include Sundays and holidays. Material sent through the postal system is not often considered to be freight.

One of the unique things about using the postal system is that the USPS will not bill the vendors for the postage. Therefore, you will virtually always be asked to prepay the estimated postage before the material is ready to mail.

The post office does not have a tracking system, and this makes it a less attractive option when shipping materials that you need to be sure were delivered. Except in the case of mass mailings, shipping via

USPS is usually not considered because the post office has so many restrictions on packages. Although the pricing is sometimes competitive, the USPS is infrequently considered when the most important issues are time and the ability to trace a package.

Package Services

Package services are carriers that offer a wide variety of delivery options ranging from next-day air to standard ground, which can take from three to ten days to arrive. Some carriers offer tracking on all packages regardless of the service you use, while others offer tracking only on air services. Weekend delivery options are determined by the carrier and the type of service you have chosen. Some carriers will not deliver ground shipments over the weekend; others do. You need to be sure that the carrier you choose has not only the best rates but also the services you need.

Costs vary as much as the various services and types of services these companies offer. Some services offer special price breaks to large customers, while others might issue special prices for any large project with many delivery addresses. We all know that the distance from the point of shipment impacts the cost of shipping a package, but not everyone is aware that delivery to a residential address can also incur an upcharge. In addition, some companies will not track residential deliveries. Size and weight limitations for standard rates can make a huge cost difference if you are shipping to multiple locations.

Air Freight

With air freight, the skids or cartons are picked up by an expediter and put on a plane for delivery to the airport nearest the final destination. The material is then taken by a ground messenger to the final destination. This is obviously very expensive, much like buying a last-minute airline ticket. There will be a stated *lead time* by which material will need to be in the hands of the expediter or shipper. Lead time is the amount of time prior to the flight required to get the material to the airport and on the plane. Once the plane arrives, there is usually a specified time allowed for unloading and checking in before the material can be transferred to ground transportation and go out for delivery. Finally there is the time required to get the material delivered by ground to the final destination. So when considering air freight, you need to know what the total amount of time required will be, not just how long a flight it is from one airport to the other.

A number of factors can delay air freight, with weather being the least predictable. Another factor is shipping weight. Depending on the carrier, the weight of the material can cause it to be bumped from the first available plane to a later one. These are the two most common delays but not the only possible reasons for late or missed deliveries. When using air freight, it is important to know what kind of guarantees the provider is offering.

Freight Lines and Truck Services

A wide range of trucking companies offer standard shipping routes into freight depots located in major cities and industrial areas on a set schedule and make local deliveries from the depots within a guaranteed amount of time. This can be an economical way to move freight because it is moving with other freight on a scheduled truck and you have a fairly tight window of delivery guaranteed.

A dedicated truck can also be arranged that will take your materials directly from the point of origin to the delivery address. If you have a particularly tight schedule, you can request a team of drivers so the truck is driven straight through, with one driver sleeping while the other drives. This is an excellent option for a heavy load even if the load does not result in a full truck because it is often much less expensive than air freight. It is also a good option in bad weather, since planes can be delayed when trucks can still get through, and since the entire truck is consigned to you, your material cannot be bumped like it can with air freight.

Appendix:
Image and Surface
Comparisons

This appendix provides a visual comparsion of various images and printing surfaces. It shows black-and-white halftones, duotones, and four-color reproductions of the same color original and other graphic elements (e.g., color bars and reversed type) printed on a variety of papers (e.g., coated, uncoated, matte, etc.). The type of paper is indicated in the lower right-hand corner of each page.

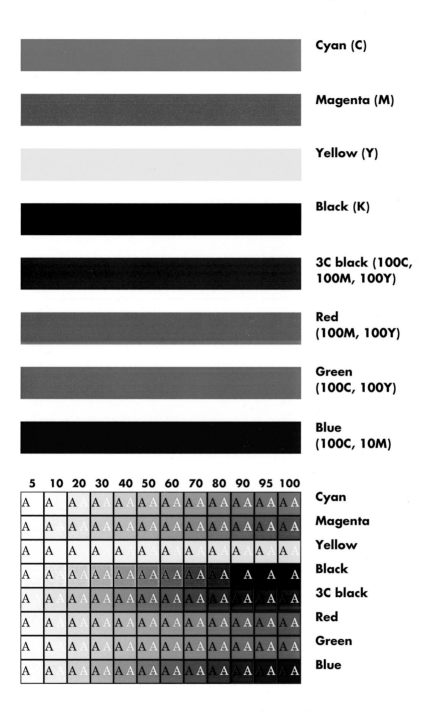

Cyan (C)

Magenta (M)

Yellow (Y)

Black (K)

3C black (100C, 100M, 100Y)

Red (100M, 100Y)

Green (100C, 100Y)

Blue (100C, 10M)

5	10	20	30	40	50	60	70	80	90	95	100	
A	A	A	A	A	A	A	A	A	A	A	A	Cyan
A	A	A	A	A	A	A	A	A	A	A	A	Magenta
A	A	A	A	A	A	A	A	A	A	A	A	Yellow
A	A	A	A	A	A	A	A	A	A	A	A	Black
A	A	A	A	A	A	A	A	A	A	A	A	3C black
A	A	A	A	A	A	A	A	A	A	A	A	Red
A	A	A	A	A	A	A	A	A	A	A	A	Green
A	A	A	A	A	A	A	A	A	A	A	A	Blue

Printed on 100-lb.
gloss coated text paper

Four-color reproduction of color original

Black-and-white reproduction of color original

Duotone reproduction of color original, printing in black and cyan

Printed on 100-lb. gloss coated text paper

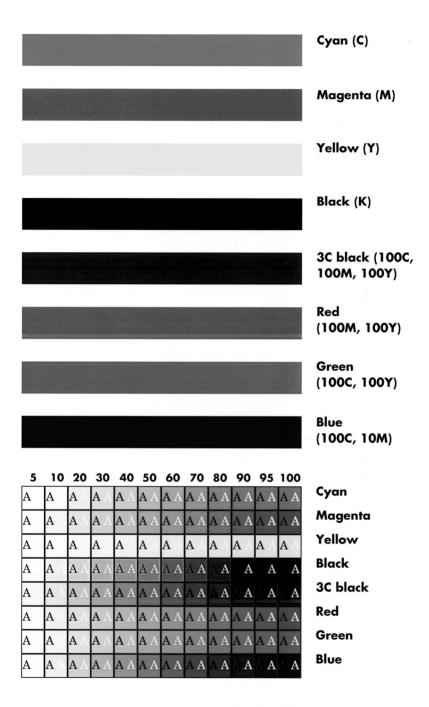

Cyan (C)

Magenta (M)

Yellow (Y)

Black (K)

3C black (100C, 100M, 100Y)

Red (100M, 100Y)

Green (100C, 100Y)

Blue (100C, 10M)

5	10	20	30	40	50	60	70	80	90	95	100	
A	A	A	A	A	A	A	A	A	A	A	A	Cyan
A	A	A	A	A	A	A	A	A	A	A	A	Magenta
A	A	A	A	A	A	A	A	A	A	A	A	Yellow
A	A	A	A	A	A	A	A	A	A	A	A	Black
A	A	A	A	A	A	A	A	A	A	A	A	3C black
A	A	A	A	A	A	A	A	A	A	A	A	Red
A	A	A	A	A	A	A	A	A	A	A	A	Green
A	A	A	A	A	A	A	A	A	A	A	A	Blue

Printed on 80-lb.
silk text paper

**Four-color
reproduction
of color original**

**Black-and-white
reproduction of
color original**

**Duotone repro-
duction of color
original, printing
in black and cyan**

Printed on 80-lb.
silk text paper

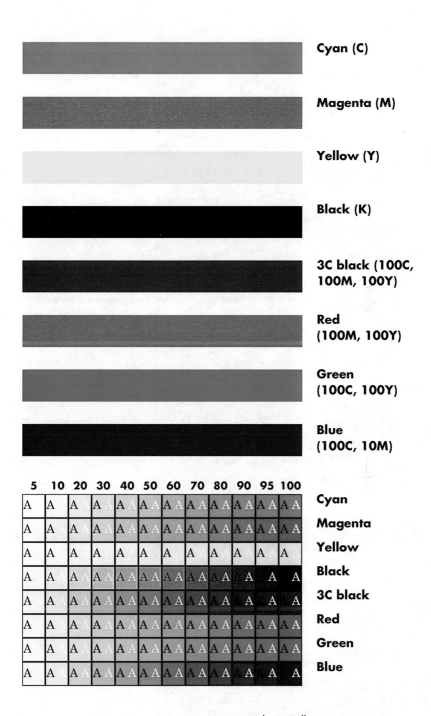

Cyan (C)

Magenta (M)

Yellow (Y)

Black (K)

3C black (100C, 100M, 100Y)

Red (100M, 100Y)

Green (100C, 100Y)

Blue (100C, 10M)

5	10	20	30	40	50	60	70	80	90	95	100	
A	A	A	A	A	A	A	A	A	A	A	A	Cyan
A	A	A	A	A	A	A	A	A	A	A	A	Magenta
A	A	A	A	A	A	A	A	A	A	A	A	Yellow
A	A	A	A	A	A	A	A	A	A	A	A	Black
A	A	A	A	A	A	A	A	A	A	A	A	3C black
A	A	A	A	A	A	A	A	A	A	A	A	Red
A	A	A	A	A	A	A	A	A	A	A	A	Green
A	A	A	A	A	A	A	A	A	A	A	A	Blue

Printed on 50-lb.
uncoated text paper

Four-color reproduction of color original

Black-and-white reproduction of color original

Duotone reproduction of color original, printing in black and cyan

Printed on 50-lb. uncoated text paper

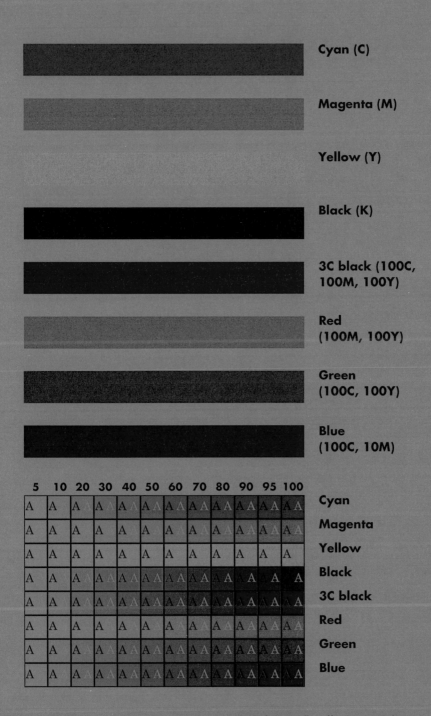

Cyan (C)

Magenta (M)

Yellow (Y)

Black (K)

3C black (100C, 100M, 100Y)

Red (100M, 100Y)

Green (100C, 100Y)

Blue (100C, 10M)

	5	10	20	30	40	50	60	70	80	90	95	100	
A	A	A	A	A	A	A	A	A	A	A	A	A	Cyan
A	A	A	A	A	A	A	A	A	A	A	A	A	Magenta
A	A	A	A	A	A	A	A	A	A	A	A		Yellow
A	A	A	A	A	A	A	A	A	A	A	A	A	Black
A	A	A	A	A	A	A	A	A	A	A	A	A	3C black
A	A	A	A	A	A	A	A	A	A	A	A	A	Red
A	A	A	A	A	A	A	A	A	A	A	A	A	Green
A	A	A	A	A	A	A	A	A	A	A	A	A	Blue

Printed on 60-lb. offset uncoated text paper (cherry color)

Four-color reproduction of color original

Black-and-white reproduction of color original

Duotone reproduction of color original, printing in black and cyan

Printed on 60-lb. offset uncoated text paper (cherry color)

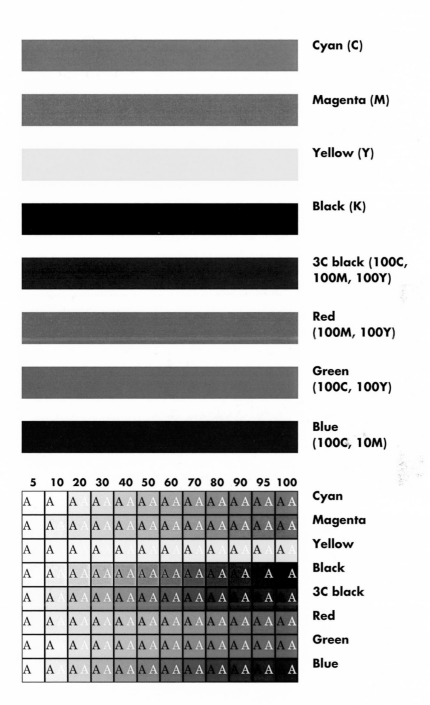

Cyan (C)

Magenta (M)

Yellow (Y)

Black (K)

3C black (100C, 100M, 100Y)

Red (100M, 100Y)

Green (100C, 100Y)

Blue (100C, 10M)

	5	10	20	30	40	50	60	70	80	90	95	100	
	A	A	A	A	A	A	A	A	A	A	A	A	Cyan
	A	A	A	A	A	A	A	A	A	A	A	A	Magenta
	A	A	A	A	A	A	A	A	A	A	A	A	Yellow
	A	A	A	A	A	A	A	A	A	A	A	A	Black
	A	A	A	A	A	A	A	A	A	A	A	A	3C black
	A	A	A	A	A	A	A	A	A	A	A	A	Red
	A	A	A	A	A	A	A	A	A	A	A	A	Green
	A	A	A	A	A	A	A	A	A	A	A	A	Blue

Printed on 80-lb. text paper (beige)

Four-color reproduction of color original

Black-and-white reproduction of color original

Duotone reproduction of color original, printing in black and cyan

Printed on 80-lb. text paper (beige)

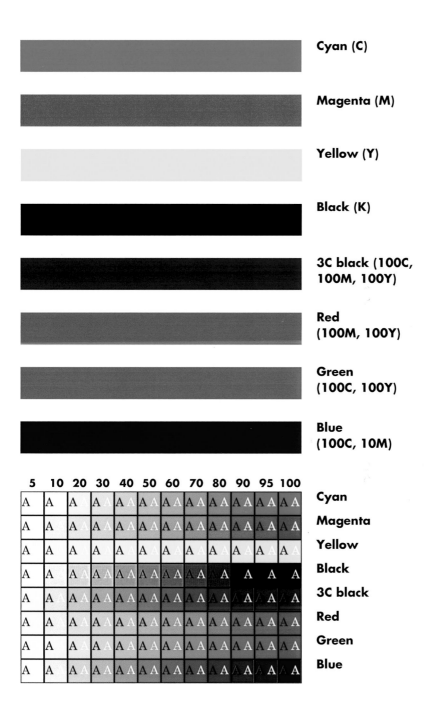

Cyan (C)

Magenta (M)

Yellow (Y)

Black (K)

3C black (100C, 100M, 100Y)

Red (100M, 100Y)

Green (100C, 100Y)

Blue (100C, 10M)

	5	10	20	30	40	50	60	70	80	90	95	100	
Cyan	A	A	A	A	A	A	A	A	A	A	A	A	
Magenta	A	A	A	A	A	A	A	A	A	A	A	A	
Yellow	A	A	A	A	A	A	A	A	A	A	A	A	
Black	A	A	A	A	A	A	A	A	A	A	A	A	
3C black	A	A	A	A	A	A	A	A	A	A	A	A	
Red	A	A	A	A	A	A	A	A	A	A	A	A	
Green	A	A	A	A	A	A	A	A	A	A	A	A	
Blue	A	A	A	A	A	A	A	A	A	A	A	A	

Printed on 65-lb. matte cover paper

Four-color reproduction of color original

Black-and-white reproduction of color original

Duotone reproduction of color original, printing in black and cyan

Printed on 65-lb. matte cover paper

Index

Index

Printing Industries of America Affiliates

Canadian Printing Industries Association
Ottawa, Ontario
Website: www.cpia-aci.ca

Graphic Arts Association
Trevose, PA
Website: www.gaa1900.com

Pacific Printing Industries Association
Portland, OR
Website: www.ppiassociation.org

Printing & Graphics Association MidAtlantic
Columbia, MD
Website: www.pgama.com

Printing and Imaging Association of MidAmerica
Dallas, TX
Website: www.piamidam.org

Printing & Imaging Association of Georgia, Inc.
Smyrna, GA
Website: www.piag.org

Printing Association of Florida, Inc.
Orlando, FL
Website: www.pafgraf.org

Printing Industries Alliance
Amherst, NY
Website: www.printnys.org

Printing Industries Association, Inc. of Arizona
Phoenix, AZ
Website: www.piaz.org

Printing Industries Association of San Diego, Inc.
San Diego, CA
Website: www.piasd.org

Printing Industries Association, Inc. of Southern California
Los Angeles, CA
Website: www.piasc.org

Printing Industries of Ohio • N.Kentucky
Westerville, OH
Website: www.pianko.org

Printing Industries of Colorado
Greenwood Village, CO
Website: www.printcolorado.org

Printing Industries of the Gulf Coast
Houston, TX
Website: www.pigc.com

Printing Industries of Michigan
Southfield, MI
Website: www.print.org

Printing Industries of New England
Southborough, MA
Website: www.pine.org

Visual Media Alliance
San Francisco, CA
Website: www.visualmediaalliance.org

Printing Industries of St. Louis
St. Louis, MO
Website: www.pistl.org

Printing Industries of the Midlands, Inc.
Urbandale, IA
Website: www.pimidlands.org

Printing Industries of Utah
W. Jordan, UT
Website: www.piofutah.com

Printing Industries of Virginia, Inc.
Ashland, VA
Website: www.piva.com

Printing Industries of Wisconsin
Pewaukee, WI
Website: www.piw.org

Printing Industry of Illinois/Indiana Association
Chicago, IL
Website: www.pii.org

Printing Industry of Minnesota, Inc.
Minneapolis, MN
Website: www.pimn.org

The Printing Industry of the Carolinas, Inc.
Charlotte, NC
Website: www.picanet.org

Printing Industry Association of the South, Inc.
Nashville, TN
Website: www.pias.org

Printing Industries Press Selected Titles

- *Adding Value to Print*, by Manfred Breede, item no. 1786

- *Customer Satisfaction Surveys: Samples from the Industry*, compiled by Printing Industries of America Human relations Department, item no. 1861

- *Customer Service in the Printing Industry*, by Richard Colbary, item no. 1594

- *Glossary of Graphic Communications*, Fourth Edition, revised by Joe Deemer, item no. 13054

- *Playbook for Selling Success in the Graphic Arts Industry: A Sales Growth Workbook for Graphic Arts Sales Professionals*, by T.J. Tedesco and Dave Clossey, item no. 1788

- *Introduction to Graphic Communication*, by Harvey Robert Levenson, Ph.D., item no. 1759

- *PrintScape: A Crash Course in Graphic Communications*, by Dr. Daniel G. Wilson, Deanna M. Gentile, and Printing Industries of America Staff, item no. 91502

- *Printing Production Management*, 2nd Edition, by Gary G. Field, item no. 1758

- *The Handbook for Digital Printing and Variable-Data Printing*, by Penny K. Bennett, Ph.D., Harvey Robert Levenson, Ph.D., and Frank J. Romano, item no. 1752

- *What the Printer Should Know About Ink*, by Dr. Nelson R. Eldred, item no. 13113

- *What the Printer Should Know About Paper*, by Lawrence A. Wilson, item no. 13083

- *Social Media Field Guide: A Resource for Graphic Communicators*, by Julie Shaffer and Mary Garnett, item no. 1790